Cambridge Monographs in African Archaeology
59
Series Editors: John Alexander and Lawrence Smith

Archaeology and Geoarchaeology of the Mukogodo Hills and Ewaso Ng'iro Plains, Central Kenya

Frederic Pearl

BAR International Series 1247
2004

Published in 2016 by
BAR Publishing, Oxford

BAR International Series 1247

Cambridge Monographs in African Archaeology 59

Archaeology and Geoarchaeology of the Mukogodo Hills and Ewaso Ng'iro Plains, Central Kenya

ISBN 978 1 84171 607 7

© F Pearl and the Publisher 2004

The author's moral rights under the 1988 UK Copyright,
Designs and Patents Act are hereby expressly asserted.

All rights reserved. No part of this work may be copied, reproduced, stored,
sold, distributed, scanned, saved in any form of digital format or transmitted
in any form digitally, without the written permission of the Publisher.

BAR Publishing is the trading name of British Archaeological Reports (Oxford) Ltd.
British Archaeological Reports was first incorporated in 1974 to publish the BAR
Series, International and British. In 1992 Hadrian Books Ltd became part of the BAR
group. This volume was originally published by Archaeopress in conjunction with
British Archaeological Reports (Oxford) Ltd / Hadrian Books Ltd, the Series principal
publisher, in 2004. This present volume is published by BAR Publishing, 2016.

Printed in England

PUBLISHING

BAR titles are available from:

        BAR Publishing
        122 Banbury Rd, Oxford, OX2 7BP, UK
EMAIL   info@barpublishing.com
PHONE +44 (0)1865 310431
FAX   +44 (0)1865 316916
        www.barpublishing.com

Abstract

    Archaeological and geoarchaeological surveys of the Mukogodo Hills and Ewaso Ng'iro Plains of Kenya are reported. Seventy-one archaeological sites dating from the Middle Stone Age to the present are described, and an alluvial chronology consisting of five sedimentary units dating from the late Pleistocene and Holocene epochs is presented. These data are used to test four hypotheses about Quaternary period human adaptation in the region.

    Hypothesis One, that landscape changes did not affect the distribution of archaeological sites in the study area, is rejected. Sites attributable to the Later Stone Age, while found on all landforms, are more abundant in the interfluvial area between the Tol and Peleta drainages near the Mukogodo foothills. The absence of Middle Stone Age sites on the Holocene surfaces in the interfluvial areas is attributable to erosion and/or burial of Pleistocene deposits containing those sites.

    Hypothesis Two, that there were no significant differences in land-use patterns between the Middle and Later Stone Age in the region, is rejected. Later Stone Age sites are clustered in the Mukogodo foothills, while Middle Stone Age sites exhibit an unpatterned distribution on Late Pleistocene surfaces. This difference is interpreted as representing land use intensification in the Later Stone Age due to population growth, circumscription, or ecological factors. Later Stone Age land-use is seen as evidence of greater strategic planning or planning depth.

    Hypothesis Three, that the arrival of pastoralism did not contribute to erosion and degradation of the landscape, could not be rejected as no direct evidence indicates that the prehistoric introduction of pastoralism into the study area had any physical impact on the landscape. Alternate hypotheses concerning the cause of erosion evident in late Holocene sedimentary deposits are suggested.

    Hypothesis Four, that changes in economic patterns in the archaeological record of the study area can be correlated with the Quaternary period environmental changes, is accepted. Significant differences in land use and lithic raw material procurement patterns exist between the Middle and Later Stone Age. These changes appear associated with the dramatic climate amelioration at the end of the Pleistocene.

"A man must judge his labours by the obstacles he has overcome and the hardships he has endured, and by these standards I am not ashamed of the results"

Evans-Pritchard 1940:9

Acknowledgements

My work in Kenya would not have been successful without the assistance of many individuals. I am most indebted to Dr. D. Bruce Dickson, Director of the Mukogodo Hills Project, who provided much needed guidance and advice over the past five years. Without his direct support, none of this would have been possible. I am also indebted to Dr. Lee Cronk of Rutgers University, who began work with the Mukogodo people of Kenya in 1986. His work inspired me to travel to Kenya ten years later to begin my own studies there, leading to my involvement with the Mukogodo Hills archaeology project. The dissertation done by Dr. G-Young Gang of Yeungnam University, Korea, as part of the Mukogodo Hills project provided a solid basis for my own later work. I also wish to thank Dr. Gang for undertaking analysis of lithic materials from the sites in the study area. I am also very grateful for the editorial support provided by Trina Arpin, Robyn Lyle and Gail Davis. The graphics were greatly improved by Elizabeth Neyland.

Since 1996, the faculty at Texas A&M University has been very helpful and inspiring. Drs. David Carlson, Fred Smeins and Michael Waters contributed directly to this work by providing commentary and editorial advice. Some of the individuals not yet mentioned, but who contributed most significantly to my intellectual growth are Drs. Vaughn M. Bryant, Norbert Dannhaeuser, Harry Shafer, D. Gentry Steele, and Alston Thoms. I would also like to acknowledge Dr. David Kuehn for his advice and guidance in and out of the field. William Seitz of Texas A&M Galveston provided encouragement, advice, and helped to procure funding that greatly improved this work—Thank you.

I would also like to thank the staff of the Archaeology Division at the National Museums of Kenya for their friendship and support. Special thanks are due to Dr. Karega-Munene, Head, and Mr. Samuel Kahinju, Collections Manager of that office. Dr. George Abungu, Director of the National Museums, was also helpful with facilitating research in Kenya. I am also greatly in the debt of the Turkana, Samburu, and Mukogodo peoples of the Mukogodo Hills and Ewaso Ng'iro plains for their guidance, loyal assistance, tolerance, and understanding.

Financial support was provided by the Memorial Student Center L.T. Jordan Institute for International Awareness at Texas A&M University. It was their initial funding in 1996 that enabled me to become involved with Africa in the first place. Additional financial support was provided by the Texas A&M University Department of Anthropology, the TAMU Program to Enhance Scholarly and Creative Activities, the TAMU Interdisciplinary Research Initiative and the Brazos Valley Museum of Natural History, Bryan, Texas. Research subsequent to 2000 was entirely supported by a grant from the Research Management Office at Texas A&M University at Galveston.

Table of Contents

INTRODUCTION 1
   The Road to Kipsing 1
   Current Archaeological Research on the Middle and Later Stone Age
   in East Africa 3
   Investigations in the Mukogodo Hills 6
   Previous Surveys in the Mukogodo Hills and Elsewhere in Kenya 7
   Outline of This Monograph 7
   Significance of the Research 8

ENVIRONMENT AND ECOLOGY 9
   The Contemporary Environment 9
   Human Ecology of Foraging 10
   Human Ecology of Pastoralism 10
   Paleoecology 13
   Pleistocene Environments 13
   Colonial-Era Ecological Trends 16

GEOARCHAEOLOGY 18
   The 1995 Tol River Profile 18
   Important Geological Processes in the Mukogodo Foothills and
      Ewaso Ng'iro Plains 18
   Kipsing Area Quaternary Geology 21
   Summary of Alluvial Chronology 24
   Predictive Site Model 24

SITE SURVEY 27
   Methods 27
   Problems Encountered 27
   Site Chronology 29
   Discussion of Survey Results 30

SUMMARY AND CONCLUSIONS 42
   Review of Goals and Hypotheses 42
   A Brief History of the Mukogodo Hills Region Based on
      Archaeological Evidence 47
   Suggestions for Further Research 50

REFERENCES CITED 52

APPENDIX A 59

LIST OF FIGURES

.... Page
Figure 1. Geographic location of Kenya on the continent of Africa 2
Figure 2. Regional map of the Mukogodo Hills and Ewaso Ng'iro Plains. 3
Figure 3. Longitudinal profile of the distance between Don Dol and Kipsing. 4
Figure 4. Oxygen isotope curve based on data in Martinson *et al.* (1987). 15
Figure 5. Locations of sedimentary profiles described in the text. 19
Figure 6. Drawing of Tol River profile based on Kuehn's 1995 notes. 20
Figure 7. Tol River profile based on 2000 interpretation. 22
Figure 8. Correlation of stratigraphic units between selected profiles 25
Figure 9. Map of the study area showing boundaries of Survey Areas A and B. 28
Figure 10. Examples of Middle Stone Age flakes from GnJm 18. 31
Figure 11. Examples of Middle Stone Age cores from GnJm 23. 32
Figure 12. Examples of Later Stone Age technology from GnJm 47. 33
Figure 13. Distribution of Middle Stone Age sites discovered on survey. 35
Figure 14. Distribution of Later Stone Age sites discovered on survey. 36
Figure 15. Distribution of Iron Age sites within the study area. 38
Figure 16. Distribution of stone cairn sites discovered on survey. 39
Figure 17. Distribution of other sites discovered on survey. 40
Figure 18. Timeline showing climate change, lake-level fluctuations,
            and cultural milestones of significance. 46

LIST OF TABLES

		Page
Table 1.	Radiocarbon Results.	20
Table 2.	Predictive Site Model.	26
Table 3.	Corpus of Archaeological Sites Recorded During the Mukogodo Hills Site Survey.	34
Table 4.	Survey Results Tabulated By Cultural Period And Quaternary Stratigraphic Unit.	42

CHAPTER 1
INTRODUCTION

Since 1996, I have been involved with a cultural, archaeological, and geological survey of the Mukogodo Hills and Ewaso Ng'iro Plains in Central Kenya (Figure 1). Results of this research are presented herein, with a primary goal of providing an environmental chronology and describing patterns of human land use through the Late Pleistocene. This was accomplished through geoarchaeological and archaeological survey. The geoarchaeological study demonstrates how local environmental conditions, particularly fluvial geomorphology, have responded to East African climatic fluctuations. This, in turn, provides a comparative basis to interpret cultural change documented by the archaeological survey. While building on research that has already been conducted in the region, these investigations provide the context within which to make archaeological interpretations meaningful.

This research addresses four main questions. First, did landscape changes affect the distribution of archaeological sites in the Mukogodo Hills-Ewaso Ng'iro Plains region? Second, are there significant differences in land-use patterns between the Middle and Later Stone Age inhabitants of the region? Third, did the arrival of pastoralism contribute to erosion and degradation of the landscape? Fourth, has ecological change correlated with changes in economic patterns observed in the archaeological record?

THE ROAD TO KIPSING

The "Mukogodo Hills" refers to the group of low hills and granitic ridges in the northern half of the Loldaika-Mukogodo Range. This range sits on the eastern edge of the Laikipia Plateau where it drops into the Ewaso Ng'iro Basin (Figure 2). This is the traditional home of the Mukogodo people who were hunter-gatherers into the middle part of the twentieth century, but who today are principally pastoralists and wage-laborers. The elevation at Don Dol is 1,840 m, while elevations of prominent nearby landmarks are up to 400 m higher. Vegetation in these uplands is thick enough to be called a "forest" by map surveyors. The Mukogodo Forest consist of cedars (*Juniperus procera*), African wild olive (*Olea africanus*), and mixed thorny acacias (*Acacia* spp.) among others (Cronk 1989a:54; and personal observation).

The Ewaso Ng'iro Basin takes its name from the Ewaso Ng'iro, the large river that drains this region. Its catchment includes over 600 square km and such towns as Nanyuki, Isiolo, Wamba, Don Dol, Archer's Post, and Kipsing. It is a perennial stream recharged by large tributaries, including some that flow from the northern buttress of Mt. Kenya. One such tributary is the Kipsing River, one of the few rectangular drainage systems in an otherwise dendritic system (Hackman et al. 1989:6–7). Its drainage network is constrained by trends of the underlying gneisses that form the Mukogodo Hills. Three ephemeral streams that drain the Mukogodo Hills region feed the Kipsing River: the Sinyai, the Tol, and the Seaku rivers. The small town of Kipsing lies where the Seaku River joins the Kipsing.

Beyond Don Dol, the road to Kipsing descends a steep escarpment at the edge of the Laikipia Plateau. The view over the Ewaso Ng'iro Basin from the Mukogodo Hills is quite spectacular. The flat valley lies about 800 m below Don Dol at an average elevation of 1000 m. At first glimpse, the valley appears to be carpeted in green. However, this is merely an illusion created by the flat tops of acacia trees that are quite dense at this end of the basin. The Seaku River drains the Mukogodo Hills, but it is dry most of the time. From the road, the Seaku River appears as a dense thicket skirting the edge of a long ridge of gneiss that extends into the desert.

Engineers who built the road looked for a more favorable grade than that taken by the Seaku. The old road winds down the face of the Laikipia Plateau, but one finds out quickly why this route was abandoned long ago. The road was built upon sediments that thinly veil the steep, rocky grade. Years of seasonal rains have wiped out the road in many locations, leaving nothing but rough bedrock. Every now and then, the trail resumes but is often cut by deep gullies. The road that descends the escarpment is treacherous on foot, but impassable by vehicles.

The route from Nanyuki to Kipsing through Don Dol was about 80 km when it was passable. These days, the few vehicles that need to get to Kipsing must travel a circuitous route either through Ol Doinyo Ng'iro to the west, or Isiolo to the east, effectively doubling the off-road distance beyond Nanyuki to about 150 km. Although, the route through Isiolo is potentially faster because part of the road is paved, the increase in lawlessness and banditry along that route in recent years makes it less attractive. Few unarmed or unescorted travelers pass that way. Most traffic heads through Ol Doinyo Ng'iro. Traveling to Ol Doinyo Ng'iro from Nanyuki means taking a splendid but rugged route that provides many spectacular views of the banded Don Dol and Ol Doinyo Ng'iro gneisses where they outcrop to the surface. These gneisses underlie the north-south trending hills that mark the northern end of the Laikipia Plateau and make for a long, hard drive. Whether you are descending from Ol Doinyo Ng'iro or from Don Dol, the flat topography of the Ewaso Ng'iro Basin is strikingly different from the uplands (Figure 3). The monotonous horizon, however, is relieved with small isolated hills of granitic gneiss (inselbergs) that attest to the subsurface geology. Inselbergs range from small pimples less than 10 m in height to rather large hills over 2 km long and 500 m tall, with names like Kakwa Lelash, Tale, and Lekupe. Far to the east towards Archer's Post and Isiolo are volcanic intrusives expressed on the surface, but nothing in the area near Kipsing.

The dominant vegetation on the plains is thorny acacia, a variety of grasses, and a few succulents. Upon approaching Kipsing, the powerful impact of grazing on the landscape is clearly evident. In the years from 1996 to 2000, when the region was in the midst of a severe drought, grass was patchy at best. More than 80 percent of the landscape was exposed in one way or another. Most of the cattle had been driven of

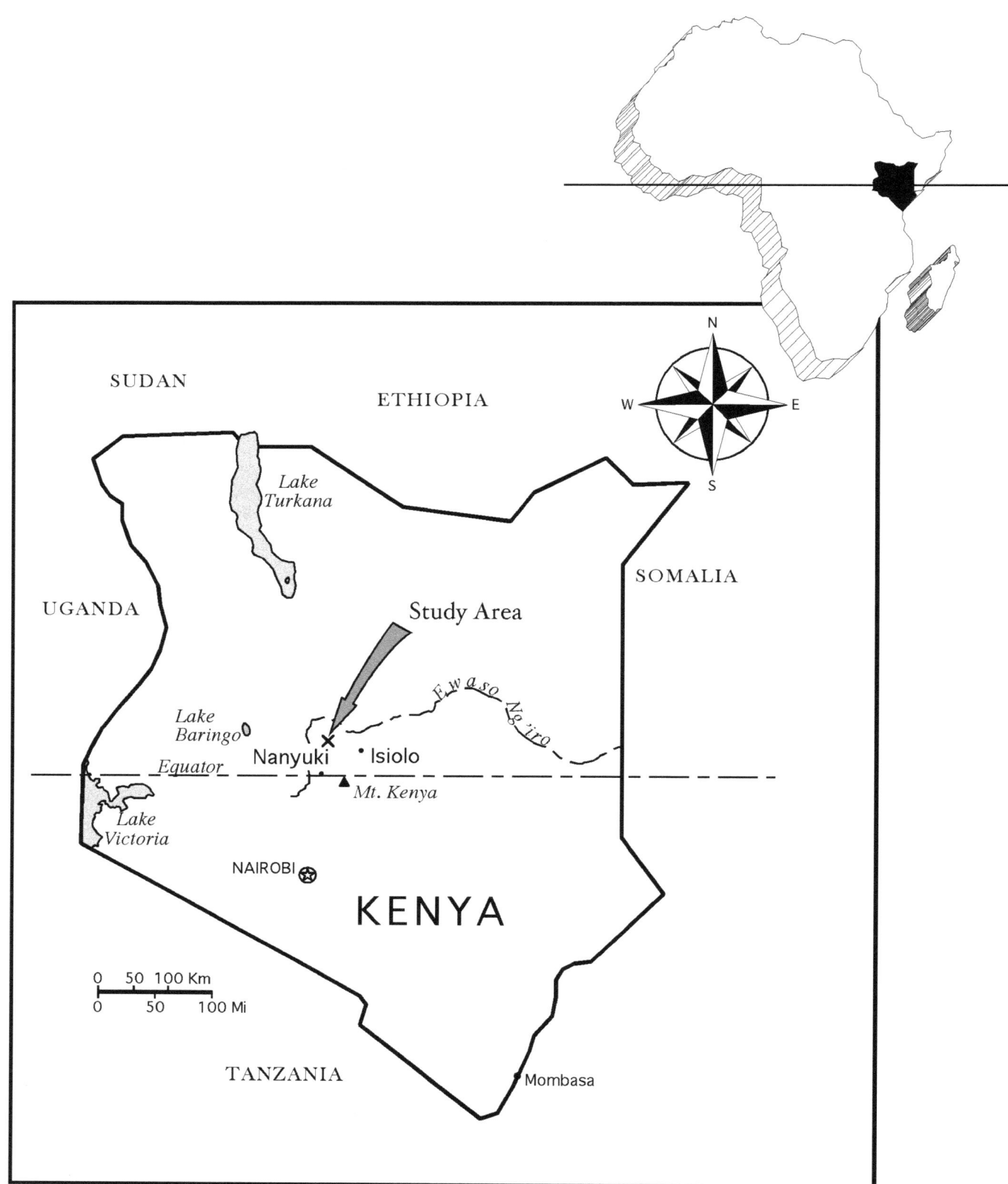

Figure 1. Geographic location of Kenya on the continent of Africa.

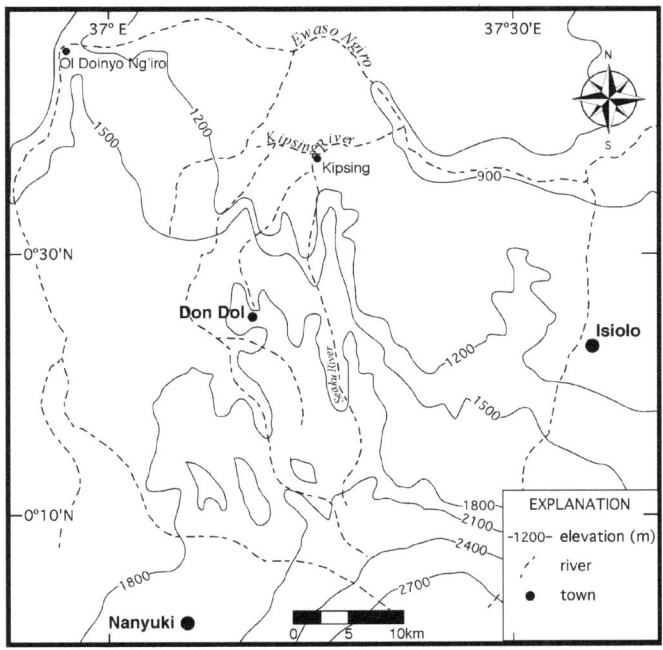

Figure 2. Regional map of the Mukogodo Hills and Ewaso Ng'iro Plains.

by their owners, leaving behind small herds of domestic goats, camels, some sheep, and donkeys. These animals are better suited to take advantage of poor quality grass, yet abundant scrub and leaf browse.

Kipsing was already a regular watering hole for Samburu cattle when the British established an outpost there during the Colonial period. Livestock enjoyed the naturally saline Kipsing waters. Kipsing continued to be an important locus of pastoralist activity after Kenyan Independence when the government established a livestock inoculation center there. In the 1980s, the Catholic Mission at Ol Doinyo Ng'iro established a satellite primary school called Kipsing Academy that principally serves the children of the region's many pastoralist families.

Ethnic violence broke out just as we arrived in the project area in 1996 as local Samburu and Somali pastoralists feuded over cattle theft. Cattle were stolen and herded right through the project area by Somali raiders. The Samburu immediately mobilized a recovery group consisting of young Samburu *moran* (younger age-set warriors), armed principally with spears. Some of these young men asked to use our project vehicle to track the cattle thieves. Our close relations with the local people obliged us to help them. Not unfortunately, however, the vehicle broke down as we were ferrying some warriors in pursuit of the Somalis. The situation failed to improve as there were more cattle thefts and related human deaths over the next few days. When the violence subsided, 13 people were dead and many of the residents who had begun to establish permanent residences near Kipsing had moved away. Today, many of the small shops stand empty near Kipsing.

Cattle raiding is problematic among East African cattle herders. Both the colonial and post-colonial governments of Kenya have tried unsuccessfully to stop this practice. By our return in 1999 the local Samburu had armed themselves with automatic weapons that were illegally imported into the region. Now, with armed Samburu *moran* here and there, and a Kenyan military contingent regularly patrolling the area, we were able to finish the work begun three years earlier.

CURRENT ARCHAEOLOGICAL RESEARCH ON THE MIDDLE AND LATER STONE AGE IN EAST AFRICA
The Early Stone Age

The Early Stone Age was first defined as archaeological period by Goodwin and van Reit Lowe in 1929. They proposed a three-age system of Early, Middle, and Later Stone Age that paralleled the European system of Lower, Middle, and Upper Paleolithic. Initially this met with resistance and was considered redundant by a few scholars (Leakey 1931). While the two systems have some overlap, assemblages in all six divisions have neither the same content nor temporal composition (McBrearty and Brooks 2000). Local terminology is often used as a subset within a larger industrial complex. Indeed, terms such as Lupemban, Sangoan, Elmentaitian, and others continue to appear as more detailed descriptions of regional artifact assemblages are discovered that still generally fit within three-age system.

The earliest archaeological evidence for the use and manufacture of tools is about 2.5 million years old (Semaw et al. 1997). The Oldowan Industrial Complex, as it is known, consists mainly of technologically simple tools created by removal of one or a few flakes to produce a sharp cutting edge (Ambrose 2001). Typical artifacts are hammers, and nonspecialized flakes and cores. Oldowan tools are always made of locally available raw materials and probably represent "expedient" technology that was created upon need and discarded upon use.

The most widespread technology that occurs in the latter portions of the Early Stone Age is the Acheulian Industrial Complex. The Acheulian industry is widespread throughout Africa, Europe, the Levant, and parts of Asia. In East Africa, it extends from about 1.5 million years ago up to about 100,000 years ago (Phillipson 1993:34). It represents a tremendous technological advance over Oldowan technology and produces the world's first widespread, standardized toolkit. It is characterized by the presence of large cutting tools, usually fashioned on a large flake or cobble. The hallmark artifact for this period is the Acheulian hand axe, a strikingly uniform bifacial axe-shaped tool. In addition to the ubiquitous hand-axe, many Acheulian sites in Africa contain other nonstandardized flake tools, trimmed around the edge, perhaps used as scrapers, borers, and the like (Phillipson 1993:36).

Another technology, the Levallois or prepared-core technique, also rose to prominence during the later Early Stone Age. Levallois technology is a sophisticated technique for producing large numbers of relatively uniform flakes (Dibble and Bar-Yosef 1995). The Levallois technique produces very characteristic flakes and cores that are easily identifiable in the archaeological record. The Levallois

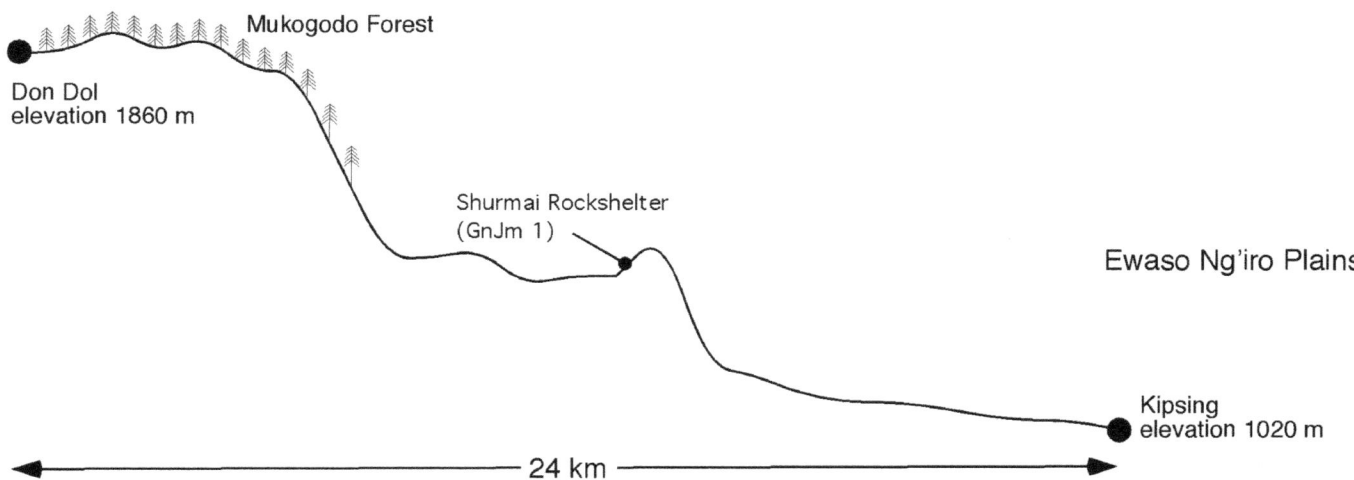

Figure 3. Longitudinal profile of the distance between Don Dol and Kipsing.

technique is also used in refined form later in the Middle Stone Age.

The Middle Stone Age

The transition from Early to Middle Stone Age has not been identified as abrupt in East Africa, but rather appears to cluster around 200,000 B.P. (Robertshaw 1995). The most recent Early Stone Age material in Kenya comes from the Kapthurin Formation, where Acheulian artifacts underlie a volcanic layer dated by the potassium-argon method to ca. 240,000 B.P. (McBrearty et al. 1996). Ash overlying Middle Stone Age material from Gademotta Hill in Ethiopia revealed a potassium-argon date of $181,000 \pm 6,000$ B.P., suggesting that the Middle Stone Age began prior to that time (Wendorf et al. 1975). In any case, the Middle Stone Age, with its toolkit of prepared cores and specialized flake and blade tools, was well underway by at least 120,000 years ago. Characteristic Middle Stone Age artifacts at Prospect Farm were dated using the obsidian hydration method to $119,646 \pm 1668$ B.P. (Michels et al. 1983).

Although formal archaeology of the Middle Stone Age (MSA) is well documented, interpreting the archaeology is one of the hottest subjects of contemporary archaeological research. The Middle Stone Age is distinguished from the Early Stone Age by the appearance of a diverse lithic toolkit that is often the product of prepared-core technology. This technology produced numerous specialized blades, backed-blades, uniform triangular flakes, Levallois flakes, and formal core types (Ambrose 2001:1751; Clark 1970:124; McBrearty and Brooks 2000:495). Also characteristic is the replacement of large cutting tools, such as hand axes and cleavers that characterized the Early Stone Age, with core axes and picks. This is particularly evident at Kalambo Falls in Zambia, where Early Stone Age and Middle Stone Age assemblages are stratified (Clark and Kleindienst 1974; Sheppard and Kleindienst 1996). These core axes and picks come to characterize the Sangoan industry of the Middle Stone Age. At Muguruk in western Kenya, Sangoan tools are found stratified beneath other, more typical Middle Stone Age tools (McBrearty 1988). Robertshaw (1995) concludes, therefore, that the Sangoan is intermediary between the Early and Middle Stone Age.

Excavations at Shurmai Rockshelter (GnJm 1) in the study area have further elucidated the technology of Middle Stone Age knappers (Gang 2001). The collection of tools at Shurmai was dominated by backed and unbacked blades, denticulated flakes, and scrapers. Cores from Shurmai are generally specialized for blade production. The lowest occupation of Shurmai was clearly delineated beneath a stratigraphic disconformity, and produced relatively crude lithics of fine-grained basalt (Kuehn and Dickson 1999:72). An infrared-stimulated luminescence measurement of feldspar grains yielded a conservative minimum age of $45,211 \pm 5,356$ B.P for the overlying lithostratigraphic unit.

Later Stone Age

The Later Stone Age is marked by several important changes that are reflected in the archaeological record. The characteristic Middle Stone Age technique of manufacturing stone tools on Levallois flakes removed from prepared cores disappears in the Later Stone Age. Instead, flakes removed from plain platform cores and subsequently broken to form microliths became the norm. A substantial increase in the diversity of artifact types and an accelerated rate of their change through time also occurred (Klein 1999:589). Of late, a significant debate has emerged in archaeology that is tied to these technological changes, as well as other inferred behavioral changes.

Some of the more prominent changes noted by archaeologists worldwide are the appearance of incontrovertible art forms, ritualized burials, formal artifacts, and possibly jewelry, made from bone, shell, and ivory artifacts (McBrearty and Brooks 2000:492). This change is so pronounced that some archaeologists have suggested that biological evolution, not merely cultural evolution, took place. In other words, biological changes in *Homo sapiens* occurred that enabled modern human behavior to emerge. This change has been called "the human revolution" (Mellars and Stringer 1989). It is compatible with mitochondrial DNA evidence that proposes a common ancestor for all humanity in Africa no more than about 200,000 years ago (Cann et al. 1987; Vigilant et al. 1991). Also called the replacement theory, this model suggests that biologically and behaviorally distinct humans originated in Africa and later migrated out and replaced the more archaic hominids then inhabiting the earth (Stringer and Andrew 1988).

The Later Stone Age is commonly thought to have begun between about 40,000 and 30,000 B.P., although some evidence suggests that a substantially earlier date may be correct. Ostrich eggshell beads radiocarbon dated to about 39,900 B.P. were discovered at Enkapune ya Muto rockshelter in the Central Rift Valley overlying Later Stone Age lithic artifacts (Ambrose 1998:388). Obsidian hydration dates from Enkapune ya Muto suggest that the transition from the Middle Stone Age occurred earlier than 46,000 B.P. The deeply stratified collection of Middle and Later layers at Enkapune ya Muto are also significant because they provide substantial documentation of changes in lithic technology and assemblage composition through time. Enkapune ya Muto is exceptional because the apparent technological transformation between Middle and Later Stone Age components is sharply defined. This is unlike some other assemblages that show a more gradual technological transition, for example Muguruk (McBrearty 1988) and Shurmai Rockshelter (Gang 2001). Though in South Africa, similar dates for the beginning of the Later Stone Age are similar to those from Enkapune ya Muto. At Border Cave in South Africa, amino acid racemization and radiocarbon dates cluster around 38,000 B.P. (Miller et al. 1993). Closer to Kenya at Mumba Rockshelter in Tanzania, amino acid racemization dates the site to about 45,000 B.P (Brooks 1996; Hare et al. 1993). These dates have led some scholars to conclude that the Later Stone Age begins in Africa about 40,000 B.P., just several thousand years earlier than it is in Europe (Klein 1999:514).

McBrearty and Brooks (2000) have made an excellent case for a much earlier beginning for the transition from the Middle to Later Stone Age in Africa, suggesting that it was more gradual and less "revolutionary" than previously thought. This debate seems to have been prompted initially by the discovery of ancient finely worked bone tools in Zaire, along the Upper Semliki River in the Western Rift Valley (Brooks et al. 1995; Yellen et al. 1995). Sites with finely worked bone artifacts, a hallmark characteristic of modern human behavior, were dated by a variety of methods to be older than 89 +22/-15 thousand years B.P. Methods employed include electron spin resonance, thermoluminescence, optically stimulated luminescence, uranium series, and amino acid racemization. Following Ambrose (1998), McBrearty and Brooks (2000:490) note that radiocarbon dates near the edge of its range (~ 40,000 years) should be considered infinite. Furthermore, the age of some other sites may have been underestimated. For example, based on uranium series dates, the earliest Later Stone Age component of Mumba Rockshelter in Tanzania may be closer to 65,000 B.P, rather than 45,000 B.P.

The significance of the debate is not just the timing of the transition, but also the nature of the transition. The commonly held "replacement" theory argues that archaic *Homo sapiens* were incapable of mastering the technologies and behavioral adaptations that fully modern *Homo sapiens* were capable of (Klein 1999:515). Wherever "modern behavior" is identified, it usually is associated with abstract thinking, planning depth, innovativeness, and symbolic behavior (McBrearty and Brooks 2000:492). If replacement has occurred, it is usually expected to be abrupt (Frayer et al. 1993). In Africa, however, change appears to have been gradual, with many of the "modern" behaviors, such as shellfishing and long-distance exchange, appearing as early as 140,000 years ago (McBrearty and Brooks 2000:530). Such evidence surely means that archaeologists must reassess any ideas about a relatively "late" replacement around 40,000 years ago. However, data do not conflict with mitochondrial DNA studies that suggest a common African ancestor about 200,000 years ago. Indeed, earlier development of modern behavior in Africa would render an early migration of humans to Australia compatible with, rather than contrary to, the replacement model (Roberts et al. 1994).

Not all scholars are ready to accept the notion of a gradual transition. Ambrose (2001:1752) notes that jewelry and finely tooled items of bone are "extraordinarily rare" in Middle Stone Age sites, and that consistent appearance of these items occurs only after 40,000 B.P. Dickson and Gang (2002) concur, at least tentatively, that changes in lithic technology in the Mukogodo Hills, notably procurement strategies and tool manufacturing and stylistic changes, is consistent with a more recent origin of modern behavior. It is clear that a great deal of discussion on this matter is forthcoming.

Later Stone Age: Pastoral Neolithic

The East African Neolithic period, characterized by pastoralism in the study area, is a time period of great interest. One of the major goals of ongoing archaeological research is to determine the timing of the onset of pastoralism in the project area. The earliest commonly accepted dates for the introduction of domesticated animal herding in Kenya come from the northern regions of the country. Sites in the Ileret area on the northeastern shore of Lake Turkana show that pastoralism was established there between 4000 and 5000 B.P. (Kuehn and Dickson 1999:72). The Pastoral Neolithic is part of the Later Stone Age, but foraging was also an important, albeit declining, economic adaptation in Kenya throughout the Holocene.

Pastoralism spread southward from the Sahara sometime after 4,500 B.P. when environmental conditions in northern Africa became dryer, and tsetse-fly breeding habitat receded westward, removing the barrier to southern expansion (Smith 1992:80). Though pastoralism is established in northern Kenya between 4,000 and 5,000 B.P., it does not spread into the well-studied southwestern reaches of Kenya for at least 1,000 years (Robertshaw 1995). This millennium lag has prompted archaeologists to propose numerous solutions (Ambrose 1998; Gifford-Gonzalez 1998; Gifford-Gonzalez 2000; Robertshaw 1995).

Analysis of faunal remains at Enkapune ya Muto centered on the record left by Holocene occupants of the cave (Marean 1992). Indeed, one of the major contributions of faunal studies in East Africa has been to help better understand what it meant to be a 'pastoralist' during the Holocene. The earliest pottery associated with a pastoralist adaptation was discovered at Enkapune ya Muto with an age of 4,860 B.P. (Ambrose 1998:381). Allowing for the standard deviation, this date is well within the range of the earliest introduction of pastoralism in Kenya. However, faunal evidence for domestication comes much later at Enkapune ya Muto. The earliest radiocarbon age estimate for domesticated faunal remains is 3,990 B.P., this indicating that the spread of pastoralism in this area was gradual, and that economic interactions between foragers and herders preceded the regional adoption of pastoralism. It is possible that pottery types represent a mixture of locally produced wares and vessels acquired through trade (Karega-Munene 1996:252–253). Certainly the occurrence of a given pottery type cannot be the sole criteria for determining the cultural affiliation of a given site.

The reason for and nature of this gradual transition is further explained by ecological factors. Recently, it has been suggested that ecological deterrents and animalian diseases (epizootic factors), particularly rinderpest, trypanosomiasis, and malignant catarrhal fever, slowed the spread of pastoralism until appropriate veterinary adaptations were developed (Gifford-Gonzalez 1998; Gifford-Gonzalez 2000). Of these, malignant catarrhal fever may have represented the most novel barrier to the spread of pastoralism in Kenya because wildebeest are its principle carriers (Gifford-Gonzalez 1998:193). It is likely that epizootic factors played a significant role in the spread of pastoralism into Kenya, just as these were factors when Europeans first arrived (see Huxley 1935:30–31). Similarly, pastoralists are expected to have established economic relations with their neighbors as part of a risk-reduction strategy, explaining the early appearance of artifacts associated with pastoralist culture among late forager assemblages (Gifford-Gonzalez 1998:194). This is consistent with other previously proposed explanations of "mixed" assemblages without domesticated faunal remains (Ambrose 1984; Bower 1991; Marean 1992). It should be noted that not all archaeologists accept the notion of gradual cultural assimilation. Robertshaw (1995) questions the validity of the earliest radiocarbon dates, and has suggested that pastoralism only arrives later, after settled agriculture makes "peripheral" pastoralism possible (Robertshaw and Collett 1983).

INVESTIGATIONS IN THE MUKOGODO HILLS

Anthropological investigations in the Mukogodo Hills region began in 1985 when Lee Cronk initiated long-term participant observation of Mukogodo Maasai life. He learned that the Mukogodo had only recently adopted the Maasai culture. At the turn of the twentieth century, the Mukogodo were foragers with a unique language (Cronk 1989a; Ehret 1995). Within only a few generations, the Mukogodo were assimilated into Maasai culture. Completion of this study prompted Cronk to return to the region with archaeologist D. Bruce Dickson to survey the Mukogodo Hills for the dry rockshelter sites that Cronk's oldest informants claimed were the principle residences of the Mukogodo foragers. In 1992 Dickson and Cronk surveyed the hilly portion of the project area for rockshelter sites (Dickson 1993), identifying 23 cave sites that showed evidence of prior human occupation. The initial goal was to reconstruct use of these caves ethnohistorically by interviewing informants who could still remember living in the caves; however, this proved impractical as even the oldest Mukogodo still alive were little children when they lived in the cave.

In a follow-up survey in 1993, Dickson located the two largest sites, Shurmai (GnJm 1) and Kakwa Lelash (GnJm 2). These two rockshelters became the locus of subsequent large-scale excavations that have provided the basis for cultural reconstruction of the Mukogodo Hills region (Dickson et al. 1998; Gang 1997). Archaeological excavation of Shurmai took place in 1993 and 1994, and excavations at Kakwa Lelash were conducted in 1995. These excavations revealed that the duration of human occupation in the Mukogodo Hills regions was long indeed. The lowest occupation of the Shurmai rockshelter was clearly delineated beneath a stratigraphic disconformity, and produced relatively crude lithics of fine-grained basalt (Kuehn and Dickson 1999:72). These artifacts included discoidal cores and moderately-sized Levallois flakes characteristic of the Middle Stone Age. No radiometric age has been assigned to artifacts from this level, though an infrared-stimulated luminescence measurement of feldspar grains yielded a conservative minimum age for the overlying lithostratigraphic unit of $45,211 \pm 5,356$ B.P. The Shurmai site, then, encompasses the transition between the Middle and Later Stone Age.

Analysis of lithics from Shurmai and Kakwa Lelash rockshelters by G-Young Gang (1997) confirmed that the entire lower assemblage contained artifacts diagnostic of the Middle Stone Age. At Kakwa Lelash, extensive deposits contained principally Later Stone Age artifacts. Dickson and Gang (2002) have evaluated the lithic assemblages of Shurmai and Kakwa Lelash rockshelters to see if any light could be shed upon the emergence of modern human behavior. They concluded that the Later Stone Age assemblage was a more sophisticated and more efficient toolkit, lending circumstantial support to the argument that "modern" human behavior is a recently developed phenomenon. At Shurmai, this change must have occurred sometime after about 45,000 years ago.

Analysis of the faunal assemblage from Shurmai Rockshelter revealed archaeology evidence of the ethnohistorically known transition of the Mukogodo from foragers to pastoralists. Mutundu (1999) suggested that the Mukogodo foragers of the late prehistoric period were specialized hunters pre-adapted to the adoption of food production. Cronk (1989b) had previously provided compelling evidence that the Mukogodo shift to pastoralism was a response to encroachment by a highly successful group—the Maasai. By entering into the pastoralist economy, the Mukogodo foragers accomplished two important objectives: 1) Mukogodo men were able to acquire wives in the new system using livestock for bridewealth, and 2) Mukogodo families were able to become larger and more economically successful within the emerging Mukogodo Maasai social system. This ultimately resulted in the near total abandonment of the foraging mode of production. Ironically, this dramatic social transformation was an apparently unintended consequence of British Colonial policy that put the Mukogodo into steady contact with foreign groups (Cronk 2000).

This research provides an important lesson. When archaeologists discover artifacts, they must consider the behaviors that produced them. Artifacts and assemblages do not evolve per se; rather, individual human beings develop or adopt new technologies in order to better achieve their objectives (Boone and Smith 1998).

PREVIOUS SURVEYS IN THE MUKOGODO HILLS AND ELSEWHERE IN KENYA

Discovering the breadth of occupation of the Mukogodo Hills region prompted Dickson to expand survey coverage to include the broad floodplain below the Mukogodo foothills. An archaeological pedestrian survey was initiated in the summer of 1996 for this purpose. Geoarchaeological investigations along the Tol River that dissects the alley between Shurmai and Kakwa Lelash revealed that a dynamic series of environmental changes occurred during the late Quaternary (Dickson and Kuehn 1997; Kuehn 1999; Kuehn et al. 1996). An extensive 4 m profile was excavated along the Tol River, and several other geological sections were also described. This study focuses on investigations begun in 1996 and the data collected during those surveys.

Since the late 1970s, a number of significant site surveys have been undertaken in Kenya (Barthelme 1985; Bower et al. 1977; Farrand et al. 1976; Foley 1980, 1981; Kusimba 1999; Kyule et al. 1997; Robertshaw et al. 1990, to name a few). Kyule and colleagues discovered an abundance of early to late Middle Stone Age and Later Stone Age sites in southern Narok District in southwestern Kenya, many with well-preserved faunal assemblages. Excavations at some of these sites have produced extensive Later Stone Age microblade industries, while fossils at other locations indicate a regional time depth of at least 2.5 million years. These sites are significant because of their great potential to reveal changes in both faunal and lithic resource exploitation strategies through time. Additional insights into the timing and nature of the Middle to Later Stone Age transition in East Africa might also be forthcoming.

The Lemek-Mara region of southwestern Kenya was particularly well documented by survey efforts reported by (Robertshaw et al. 1990). This survey resulted in the discovery of about 150 Later Stone Age or younger sites. All exhibited a surface expression and no rockshelters were discovered. The fluvial geomorphology of the Lemek-Mara area, particularly the Lemek and Oldorotua systems where the bulk of the sites were found, appears similar to the Tol and Seaku drainages in our project area. Both areas are dominated at the surface by Holocene deposition. The Lemek-Mara survey demonstrates the difficulty in locating Middle Stone Age or older surface sites in a dynamic geomorphic environment. This makes interpretation of Middle Stone Age cultural resource use problematic (Robertshaw 1995:60). The Lemek-Mara survey established the utility of obsidian studies for to establish local chronologies and to aid in the interpretation of economic behavior (Merrick et al. 1990; Michels 1990).

Foley's (1981) survey of the Amboseli basin is particularly interesting from a behavioral-ecological standpoint because he attempted to reconcile artifact distribution with ecological variables. (Foley 1981). Foley's analysis concentrated on individual artifacts as the unit of analysis and comparison across the entire project area, a procedure he calls "off-site archaeology" (Foley 1981:31). That ingenious approach, however, has not caught on in Kenya, perhaps because it is very labor intensive, the data are not easily understood, and the results are not easily compared with other surveys.

OUTLINE OF THIS MONOGRAPH

First and foremost, this study is scientific positivist research. I assume that "there is a real, knowable (empirically observable), orderly world" (Watson et al. 1971:22). Furthermore, I believe the best way to learn about this world is through the scientific method. This method, among other things, involves developing testable hypotheses and collecting independent data to test them. These hypotheses need not be complicated, but they must be able to be falsified (Popper 1959 [1934]). One of the first challenges of research is to determine hypotheses about the archaeological record that can be tested, and to determine which data that are needed to test them. Here are the null hypotheses evaluated in the current study:

1. Landscape change (patterns of erosion, deposition, and stability) has not affected the distribution of sites in the Mukogodo Hills-Ewaso Ng'iro Plains region.
2. There is no significant difference in land-use patterns between Middle Stone Age and Later Stone Age inhabitants of the Mukogodo Hills-Ewaso Ng'iro Plains.
3. The arrival of pastoralism did not contribute to erosion and degradation of the landscape.
4. Ecological change resulting from climate change is not correlated with changes in economic patterns observed in the archaeological record.

The theme of these hypotheses is cultural ecology. A basic premise of this research is that human beings are part of the natural ecosystem. As such, it is incumbent upon the archaeologists to evaluate the entire ecological system in order to understand the human component. This is consistent with a general systems approach, and key to the human behavioral ecology program to which I subscribe. Human behavioral ecology (or evolutionary ecology) is the application of evolutionary theory to the study of human biological and cultural adaptation within an ecosystem (Winterhalder and Smith 1992:5). Sections of this publication are arranged to flow from environment to culture.

This chapter introduces the physiography and cultural geography of the study area, as well as outlines goals of the current research and results of previous research. The second chapter, *Environment and Human Ecology*, describes the climate and environment of the study area in greater detail. The first part of that chapter places the environment within the context of human ecology of foraging and pastoralism – and by extension to the archaeology of the area. The latter part is dedicated to presenting the environmental data relevant to East African paleoclimates. The third chapter, *Geoarchaeology*, presents results of geoarchaeological investigations. It describes the alluvial chronology of the study area and the methods used in making those determinations. The fourth chapter, *Site Survey*, presents results of the archaeological survey, where 71 new archaeological sites were recorded over approximately 24 square km. The final chapter summarizes and evaluates data presented in other chapters.

SIGNIFICANCE OF THE RESEARCH

Scientific understanding of human prehistory and the ancient environment has been transformed in a truly revolutionary manner over the last 30 years. It is fair to say that geological and archaeological research in Africa has been at the center of this transformation. It is now clear that the eastern and southern portions of that continent contain a continuous and detailed record of environmental change and human biological and cultural evolution extending back at least six million years. Much of the data upon which this record is based have been recovered in Kenya. Yet, from a scientific standpoint, that remarkable East African nation still remains largely unexamined. Our study area in eastern Laikipia District is a case in point. Although thought to be immensely rich in cultural resources, eastern Laikipia remained an archaeological blank spot until the summer of 1992, when, in concert with the National Museums of Kenya and the Department of History of the University of Nairobi, Dickson began a long-term field study there. During the course of this field study, we have begun basic geological, archaeological, and paleoenvironmental research that promises to fill in this blank spot and expand our understanding of human origins in Africa.

Archaeological sites of the Mukogodo Hills region and the Ewaso Ng'iro Plains have the potential to yield valuable insights into the most significant problems that are currently being debated by Kenyan archaeologists investigating Middle Stone Age and later sites. The time span known to be represented in the dry cave sediments of Shurmai (GnJm 1), Kakwa Lelash (GnJm 2), and presumably some other uninvestigated rockshelters, encompasses the transition from Middle Stone Age to Late Stone Age to Iron Age, as well as the ethnographically documented transition from foraging to pastoralism. Since nearly all previous research on these transitions has come from studies in the Central Rift Valley, the Mukogodo Hills region provides an ideal testing ground for theories regarding this transition. It will be interesting to note whether or not pastoralism reached the Mukogodo Hills prior to the more southerly locations near Lake Naivasha. Furthermore, additional evidence regarding the nature of the transition to pastoralism is surely forthcoming from this region.

CHAPTER 2
ENVIRONMENT AND ECOLOGY

Over 60 years ago the anthropologist Julian Steward laid out the basic premise of cultural ecology. According to Steward (Steward 1938:260), "It is difficult to see how in any society the extent and force of purely cultural and psychological determinants can be ascertained if the ecology which conditions and delimits them is unknown." Since then the cultural ecological approach has become very sophisticated, but its basic premise still holds true. Broadly speaking, cultural ecology is the study of how humans have adapted to the natural resources of the environment and to other human groups (Campbell 1995:7). A similar definition has been employed by the contemporary cultural ecologist Napoleon Chagnon during his research on the Yanomamö of Venezuela and Brazil (Chagnon 1997:45). Chagnon, like other human behavioral ecologists, notes that the proper unit of analysis is individual adaptation to physical and social environments. This is the preferred method of applying cultural ecological principles, usually measured in terms of inclusive fitness. However, since individual behavior is largely invisible in the archaeological record, it is very difficult for archaeologists to maintain a strictly human behavioral ecology approach.

This difficulty is reflected in the contemporary archaeological debate between evolutionary archaeologists and evolutionary ecologists. Evolutionary archaeologists suggest that because individual behavior is archaeologically invisible, the focus of archaeological interpretation should be upon artifact assemblages as phenotypic variations that are subject to natural selection. The implication is that artifactual change is evolutionary change (Lyman and O'Brien 1998:616). Evolutionary ecologists have criticized this approach in that it ignores the complexities of cultural transmission (Boone and Smith 1998:S157). Simply put, evolutionary ecologists argue that the focus of archaeological investigations should be the behaviors that produce artifact assemblages. Hypotheses about the archaeological record should be formulated with the understanding that individuals use and adapt their ideologies, social structures, behaviors, and material cultures to respond to dynamic environmental conditions. Taken broadly, the environment includes physical, social, and technological aspects. In accord with this philosophy, the most parsimonious explanations of changes in the archaeological record demonstrate that the cultural change in question was a product of systemic ecological changes acting on individual inclusive fitness. Given the primacy of the environment in the scheme of cultural ecology, the prehistoric ecology or paleoecology of the study area is presented below, followed by a discussion of human foraging patterns.

THE CONTEMPORARY ENVIRONMENT

Mukogodo Hills and Ewaso Ng'iro Plains are best characterized as arid to semiarid environments. Yearly rainfall in the Mukogodo Hills between 1964 and 1984 ranged from 223.1 mm to 1048.6 mm (Cronk 1989a:52). Rainfall over the more arid Ewaso Ng'iro Plains to the north was substantially less, with rivers such as the Tol and Seaku dependent upon upland rainfall. Rainfall distribution is sporadic, with many months marked by no rainfall at all. Generally speaking, rainfall tends to have a bimodal distribution clustered around April and November. Temperatures are less variable, with a consistent daily maximum between about 30°- 34° degrees C. Daily temperature variations can be quite extreme, however, with nightly temperatures routinely dropping into the upper teens.

In a continuum that includes grassland, open woodland, and dense woodland, the Mukogodo Hills should be considered dense woodland. Much of the Mukogodo Hills is gazetted as "Mukogodo Forest." The canopy is dominated by *Acacia* spp., most often *A. tortilis*. It favors a more arid environment and is often associated with *A. abyssinia*. Also called the umbrella tree or umbrella thorn, *A. tortilis* thrive in hot, arid climates where rainfall is up to 1,000 mm per annum. However, it is extremely drought tolerant and survives in climates with less than 100 mm annual rainfall with long, erratic dry seasons (Duke 1984). *A. tortilis* is usually 4 to 15 m tall, often with several trunks. Under extremely arid conditions, it is sometimes reduced to a small wiry shrub less than 1 m tall. It produces aromatic white flowers in small clusters and flat seed pods that sustain domesticated and wild browsers, such as goats, camels, and giraffes. Notably, many species of acacia have thorns that sometimes exceed 7 cm in length. Pastoralists often use thorny acacia branches to makes livestock enclosures. In addition to acacia, the Mukogodo Forest has noticeable communities of juniper (*Juniperus procera*), African wild olive (*Olea africanus*), and another thorny species, *Commiphora* sp.

The Ewaso Ng'iro Plains, on the other hand, is much less dense in terms of tree cover. It is more properly classified an open woodland or wooded grassland. The same tree species that are found in the forest are also found on the plains, especially *A. tortilis*. Interspersed with the woody vegetation of the plains are grasses. One major type is *Themeda triandra*, or red oat grass. *Themeda* can reach almost 1.5 m in height and is easily recognizable by its reddish flowering heads. It is the most widely distributed grass in East Africa (Lind and Morrisson 1974:88). Its ground cover, however, is not necessarily dense and is often associated with a good proportion of bare ground. In all areas where *Themeda* is the dominant grass species, grass fires are a natural part of the ecosystem. Since *Themeda* owes much of its success to grassland-fire ecology, it can disappear if grazing is intensified to a degree sufficient to prevent wildfires (Lind and Morrisson 1974).

Different animal species prefer different habitat types. For example, species such as Grant's gazelle and wildebeest prefer open grassland, but they are also found in open woodland. Some species prefer more cover; for example buffalo, giraffe, and impala, are more common in an open woodland environment. Other species, such as lesser kudu and leopard, prefer dense woodland. Of course, there is a tremendous degree of overlap and many species are found in multiple environments (Lind and Morrisson 1974:99). In the current study area, elephants, elands, impala, and dik dik have all been observed in both the open woodland environment of the Ewaso Ng'iro Plains and the more dense woodland of the Mukogodo Forest. Grant's gazelle and zebra are more exclusively found on the plains, but are not entirely absent from the uplands. Since water is more abundant in the uplands, in times of drought, humans, livestock, and many wild animal species head for the hills.

HUMAN ECOLOGY OF FORAGING

Ecology is the study of the relationship between a species and its environment (Campbell 1995:7). Normally, species are studied as components of an ecosystem, which is a natural association of living creatures and inorganic substances that interact to change matter. The current study can be properly classified as human ecology since the species of interest is *Homo sapiens*. While every living member of an ecosystem exerts pressure on some other set of system components, prior to pastoralism humans probably played a relatively minor role in shaping the physical environment. We can hypothesize that prior to the introduction of domestic livestock into the region, wild herbivores such as gazelle, wildebeest, and impala were abundant. Indeed, analysis of faunal remains from Shurmai Rockshelter (GnJm 1) in the study area shows that some species, such as *Sincerus* (cape buffalo), were present where they are now absent (Mutundu 1999:30). Perhaps more importantly from an environmental perspective, populations of megaherbivores, especially elephant, rhinoceros, and giraffe, would have been more numerous prior to pastoralism. Owen-Smith (1988) has proposed that herbivore species composition is critical in shaping an ecosystem. Megaherbivores help maintain an open, diverse habitat by disrupting the growth of woody plant species. Studies from Africa indicate that when elephant populations are allowed to proliferate, populations of *Acacia* and other common African trees become dramatically reduced by trampling, uprooting, and debarking (Owen-Smith 1988:226-227). In some cases elephants have contributed to the complete conversion of woodlands to open grasslands (Bourliere 1965; Owen-Smith 1988:231). In other cases where deforestation is caused principally by fire, elephants prevent the regrowth of woody plant species and thereby maintain the open environment. Similar, though less drastic, grassland maintenance contributions are made by giraffes and rhinoceros.

The end result of a megaherbivore-maintained environment is a highly productive open grassland environment. Populations of small and medium-sized herbivores are able to flourish, and therefore populations of carnivore species, including scavenging birds, will also increase. It can be inferred at this point that an omnivorous species, especially one that would thrive in an environment rich in small and medium game, would also thrive (Owen-Smith 1988:294). Prior to the introduction of pastoralism the environment was well suited for that ubiquitous omnivore – the human species.

Pastoralism has been very successful at supplanting foragers throughout Africa in areas capable of supporting cattle. This has excluded the more humid areas where the tsetse fly thrives, as well as exceptionally arid regions where neither forage nor water are available in sufficient quantities to support a full pastoral adaptation. As a result, there are no contemporary foraging groups occupying ecological zones exactly like the Mukogodo Hills region prior to pastoralism for comparative purposes. The Hadzabe (Hadza) who live on the eastern shores of Lake Eyasi in Tanzania, occupy a similar semiarid savanna ecosystem in the Rift Valley. The most significant difference is the infestation of the Lake Eyasi region with tsetse fly, which poses a significant barrier to pastoral expansion. In that region, elephant, rhinoceros, giraffe, eland, zebra, wildebeest, hartebeest, waterbuck, impala, Thompson's gazelle, warthog, baboon, lion, leopard, and many smaller animals are common (Woodburn 1972:50). Interestingly, the one animal that the Hadza do not hunt is elephant. Woodburn (1972:52) suggests that this is due to the weakness of the poison that the Hadza use to fell large prey. Perhaps an unintended side effect of not hunting elephant is that the foraging productivity for other wildlife actually increases. The Hadza never camp on the open plains, but prefer to make camp in the trees or near rock outcrops (Woodburn 1972:50). Similarly, the Mukogodo foragers typically occupied the uplands, especially rockshelters, but foraged widely (Cronk 1989a:55). The Mukogodo generally did not hunt elephant for meat, but instead for ivory (Cronk 1989a:58-59.)

The pre-pastoralist human impact on the environment was probably less significant than that of the animals they hunted, especially megaherbivores. At this time it is unclear whether any African hunters directly caused the extinction of any species, as may have happened on other continents, though human predation may have been partly responsible for the extinction of the giant hartebeest and giant buffalo at the end of the Pleistocene (Klein 1984). It is likely that the co-evolution of megafauna and man in Africa over many millennia has produced an ecological equilibrium that was not achieved elsewhere.

Understanding human ecology—the ecosystem and the place of humans within it—is a very important part of developing models for culture change through time. This method of modeling the past has been effectively used by archaeologists since the 1960s and continues to be an important part of contemporary archaeological theory. Below, prehistoric land-use patterns in the Mukogodo Hills and Ewaso Ng'iro Plains are examined. Observations made in this chapter form the theoretical platform upon which a model for prehistoric cultural change will be based.

HUMAN ECOLOGY OF PASTORALISM

Pastoralism is a very effective subsistence strategy for supporting moderately large human population densities when the growing season or other environmental constraints preclude reliance upon farming. In the Mukogodo Hills region pastoralism is the predominant subsistence strategy. The following constraints affect the human populations living in the Kipsing area:
- Economic opportunities in the market economy are poor, making self-reliance a necessity.
- Population densities are moderate.
- The growing season is very short.

- Infrastructural development, in terms of roads, electricity, irrigation, and telecommunications, is rare to nonexistent.
- Residents are unable to legally hunt many game animals.
- Dry goods, especially cereal grains and sugars, are available at local markets.
- Livestock inoculations and auctions are made possible by the local government.

The Kipsing area is predominately inhabited by Samburu pastoralists. There are several other ethnic minorities, the largest being Turkana who, since 1993, have formed a small close-knit community. Since 1996 when ethnic fighting started to occur, the Turkana have nucleated their settlement near the Kipsing primary school and store. The Samburus normally outnumber the Turkana by at least nine to one, but this ratio may drastically fluctuate depending upon the availability of pasture. As recently as 1996 there were many Somali pastoralists living here as well, but they have since moved eastward. A fraction of the inhabitants near Kipsing are of other ethnic minorities. Most are working here as teachers or administrators at the primary school. Since 1997 there has been a small contingent of Kenyan military personnel stationed near Kipsing.

Economic opportunities for individuals living in the area are severely limited. Both men and women can find an occasional wage-paying job, but for the most part it is short-term low-paying work. In 2000, one day of human labor could be purchased for between $1.25 and $4.00 (USD). Semi-permanent work is sometimes found working for the government, as a teacher, or small businessman, but most wage-labor is opportunistic. The community does have some residents who have found relatively well-paying jobs, but these are few in number and are typically located some distance away. Small hamlets like Kipsing or Don Dol experience out-migration as workers head to urban areas for wage jobs. For those who remain, self-reliance is an important risk reduction strategy, and most families living in the Mukogodo Hills and Ewaso Ng'iro Plains are food producers.

Foraging as a primary mode of subsistence is not a reasonable option for most area residents for a number of important reasons. Pastoralism supports human populations in far greater numbers than foraging, but has very dramatic consequences on the ecosystem, which ultimately works against the foraging groups. Pastoralists and foragers find themselves in critical competition for resources. This not only includes land-use rights, but social resources as well, such as potential marriage partners. Even if foragers could persist by retreating to lands unsuitable for pastoral use, they would not have been able to intermarry with pastoral groups without fulfilling the necessary bridewealth obligations common in the pastoral systems. Cronk (1989b) has demonstrated that this indeed was an important factor in explaining why the Mukogodo foragers slowly assimilated with Maasai immigrants. By obtaining livestock, the Mukogodo foragers accomplished two important objectives: 1) Mukogodo men were able to acquire wives, and 2) Mukogodo families were able to become larger and more economically successful within the emerging Mukogodo Maasai social system. The price of this transition is evident today in the Mukogodo area, where to the casual observer there is no evidence of the older Mukogodo forager culture. The Mukogodo language (Yaaku), attire, material culture, and other traditions have all been replaced by Maasai forms. The transition to pastoralism is made irreversible due to population growth, while ecological changes that accompany pastoralism make foraging less productive.

The land of Laikipia, including the Mukogodo Hills region, was considered excellent pasture by the British, who relocated the Maasai from the Rift Valley to Laikipia after 1904. Even though this land was coveted by the British for ranching, the altitude is high (1,000—2,000 m) and the growing season is short (Ward and White 1971:107). Cultivation is out of the question without substantial investment in infrastructural development. In some localized areas this has been accomplished, but for the bulk of the region and for all of the study area, there is very little infrastructural development. Some pastoral families in the cool uplands attempt to supplement their herds with small-scale gardens, but this option is not available to families in the arid lowlands and any sort of farming is out of the question. Given these constraints, pastoralism is the most viable subsistence strategy. The government has improved access to the Mukogodo hills by maintaining a dirt road, and occasional veterinary services are provided to the pastoralists at a minimal charge. Access to Kipsing is considerably more difficult and residents must travel far to obtain veterinary services (though some medicines can be purchased from traders). Pastoralism is profitable as pastoralists are able to sell some of their livestock for cash, which can then be used to purchase manufactured goods and dry foodstuffs. This has not led to large-scale capitalist livestock production by the pastoralists; they continue to find the social value of livestock for marriage transactions and the stored nutritional value of blood, milk, and meat to outweigh the market value.

Pastoralism affects the local ecology in a number of ways:
- It increases the absolute numbers and proportions of domesticated animals.
- It decreases absolute vegetation cover.
- It contributes to land degradation due to heavy grazing land use.
- It potentially contributes to desertification under certain conditions that preclude pastoralist mobility.
- It potentially contributes to grassland expansion due to deliberate land-management attempts (e.g. fires or deforestation).
- It increases plant biodiversity as part of the same process.
- It decreases the occurrence of animals that reduce the productivity of pastoralism (e.g., predatory cats).
- It increases the fertility of local soils due to the increased input of manure.
- It decreases the proportions of other ungulates and grazers due to increased competition for grazing opportunities.
- It places increased demands on natural resources such as wood for fires and water.

The effects of pastoralism are well studied—the absolute numbers and proportions of domesticated animals in relation to wild animals are increased, vegetation cover is decreased, and environmental change ensues due to heavy grazing. Pastoralists have been known to start grass fires from time to time in order to increase primary productivity (Baker 1975; Evans-Pritchard 1940; Jacobs 1965). However, there is some debate over the extent to which land degradation will occur under free-ranging pastoral use. Use of the term "desertification," which implies irreversible degradation with advancing deserts caused by overgrazing, has fallen into disfavor because the link between grazing and irreversible land degradation has not been adequately demonstrated. The notion that communal grazing would ultimately lead to land degradation was made popular by Hardin (1968) in his proposition of the tragedy of the commons. Under this theory, communal land use inevitably leads to degradation because the benefits of adding additional animals to a herd are fully gained by the owner, but the costs of that additional animal on the environment are born by the entire community. This will ultimately lead each individual owner acting in self interest to exceed the grazing carrying capacity through overstocking individual herds.

There is no compelling evidence that desert areas are expanding (desertification). Outside of settled areas, there is little or no evidence that severe degradation is taking place (Dodd 1994; McCabe 1989). Furthermore, it has been argued that degradation that is taking place has more to do with the contemporary provisioning of education, health care, and religion than with pastoralism (Swift et al. 1996). This is true near Kipsing where it is estimated that up to 80 percent of the ground cover had been removed by grazing near the settled area that includes an educational facility, a medical dispensary, and a church. Similar ecological effects were noted north of the study area in south Turkanaland (Ellis and Swift 1988). In Ngisonyoka, Turkana District, land that appeared severely degraded during drought recovered quickly after rainfall (Swift et al. 1996:264). Thus, what ecologists and anthropologists suggest is that the contemporary ecological situation of pastoralist areas has little to do with pastoralism, and everything to do with the unintended consequences of government policies.

Both the colonial and post-colonial administrations of Kenya have supported the conversion of suitable pastoral land into ranch and agricultural land. Consequently, pastoralists can no longer seasonally migrate and are forced to put limited pasture into full-time use. Additionally, human and cattle population densities have risen in the pastoralist areas, partly due to enhanced medical and veterinary care, but also due to forced migrations that occurred in the last century. The human ecology of pastoralism since colonization is thus dramatically different from pre-colonial pastoralism.

There are also lesser-known effects of grazing on the ecosystem. The same process that reduces vegetation may also increase biodiversity. It has been suggested that pastoral use may create a patchy environment with more habitats than if it had been left ungrazed (Swift et al. 1996:263). Researchers in Laikipia documented highly productive microhabitats that formed on the abandoned sites of pastoralists settlements (Young et al. 1995). These glades have highly fertile soils undoubtedly due to the input of domesticated animal manure. Each glade supports many plant species not found elsewhere, and wild mammals also find these glades attractive. These glades represent "a relatively permanent community mosaic that increases ecosystem heterogeneity and resource use by domestic and wild animals" (Swift et al. 1995:97).

Pastoralists find themselves in indirect competition with many of the other wild herbivores that depend on available forage. Swift (*et al.* 1996:262) note that in northwest Kenya and Ethiopia wild herbivores coexist with pastoralists but in highly reduced numbers, and generally distribute themselves in such a way as to avoid contact with humans or their livestock. Wild carnivores that might prey on livestock are regarded by pastoralists as 'the enemy' and are vigorously persecuted when encountered (c.f. Swift et al. 1996:263).

Herd composition also affects the local ecology. During the extended drought from 1996—2000, though hand-dug well water was available near Kipsing in the Kipsing River, most pastoralists had driven their cattle to locations where water was more abundant. But much of the pastoralist population near Kipsing is poor, even by pastoralist standards. These pastoralists ad few or no cattle, though they had sometimes sizeable numbers of small stock, principally goats. When the cattle were moved away some herders remained behind with their goats and camels. Camels and goats are able to subsist on browse (shrubs and trees), while cattle are unable to do so. Both can eat acacia, and camels seem to be unaffected by the thorny branches (Smith 1992:107). At Kipsing, some herders have become sedentary or semi-sedentary, and the traditional pattern of transhumance is not in effect. Land is under pastoral production year round, though livestock types are informally rotated. It is uncertain what the effects of this intensive pastoralism are on the local ecology, but this novel problem is undoubtedly due to increasing population densities, restriction of pastoralists to smaller ranges, and the wider availability of markets in which pastoralists can sell livestock and purchase manufactured goods and supplemental foods. These material factors have preceded the permanent settlement of pastoralists near Kipsing and the consequent ideological changes that accompany this new way of life.

The combined effects of the proliferation of pastoralism in East Africa can thus be summarized: As a subsistence pattern, pastoralism significantly increases the human carrying capacity of the semiarid environment. After the introduction of pastoralism into the ecosystem, human populations grow and their impact on the environment and their importance in the ecosystem dramatically increases. The most overwhelming impact, however, is related to livestock upon the environment. Plant species tolerant of intense grazing become more abundant. The impact of pastoralism is unevenly distributed, however, and a patchy but diverse environment is the result as some areas are in use, others are recovering, and still others are evolving into new microhabitats due to the localized input of manure (Young et al. 1995). Biological diversity may remain unchanged, but absolute numbers of herbivores and carnivores decrease because of competition for resources with humans and their livestock. Eventual overgrazing can lead to increased erosion. In the lowlands, this has resulted in a sparse acacia savanna that supports moderate densities of wild animals, domesticated animals, and humans.

PALEOECOLOGY

Paleoecology, or prehistoric ecology, relies upon study of prehistoric rainfall patterns, temperatures, and wildlife and vegetation regimes. Fortunately, though there is very little direct evidence of any of these things in East Africa, such information can be inferred from proxy measures. Knowledge of East African paleoclimates comes principally from four sources: pollen records, lake levels, glacial evidence, and oxygen isotopes (Livingstone 1996).

Pollen records provide excellent evidence for prehistoric biome composition; however, variations in pollen production by species and differential preservation of pollen between and within sites makes this record partial at best, and at worst may lead to erroneous inferences. Since individual vegetation taxa can serve as indicators for both temperature, rainfall, and human action, even broken pollen records can lend important information about the past. Pollen records are only available for the Late Quaternary, limiting their use in earlier paleoenvironmental reconstruction.

Lake levels are indicators of the regional moisture budget (precipitation versus evaporation, or P-E ratio). Since the P-E ratio can change due to changes in rainfall and/or changes in aridity, interpreting paleolake levels is complicated. Lake levels might change independently of temperature changes as well, though clearly glacial accumulation at both the regional and global scale can affect the local moisture budget. A more thorough understanding of paleoclimates and paleoenvironments will come from an interpretation of the systematic nature of the environment, and from a composite picture using all available data.

Glacial evidence not only consists of geomorphic evidence, such as moraines and other physical features, but also relies upon isotopic data continued within the ice itself. Oxygen isotopes, which are temperature dependent, have been very helpful in determining the air temperatures under which ancient glacial ice formed. This, in turn, has allowed scientists to reconstruct temperatures into the ancient past.

Broad patterns of global climate change can be understood by examining the record of oxygen-isotopes contained in deep-sea sediment cores. This type of research calculates the ratio of isotopically heavy oxygen (^{18}O) to the lighter variety (^{16}O). Originally, this ratio was predicated on the assumption that $\delta18O$ variations were due to changes in ambient sea temperatures (Emiliani 1955). Since then it generally has been agreed upon that $\delta18O$ variations are in fact caused by changes in global ice-sheet volume (Shackleton 1967; Shackleton and Opdyke 1976). However, though ice volume and subsequent lowering of sea levels may be the dominant cause of $\delta18O$ variations, researchers recognize a host of other problematic factors that complicate interpretation (Imbrie et al. 1984:270). Fortunately these factors have been identified and incorporated into the widely accepted models used here (Imbrie et al. 1984; Martinson et al. 1987). Isotopic events labeled 2, 3, etc., are as described by Shackleton and Opdyke (1976).

By correlating the peaks and valleys of the oxygen-isotope curves with regional glacial advances and lake-level fluctuations, the effect that changes in global ice volume has on the regional moisture budget becomes clear. In turn, changes in temperature and moisture coincide with changes in vegetation, which can be seen in pollen studies. Animal populations undoubtedly also changed as part of the overall biome transformations that characterize the past. Wildlife and vegetation are of immediate concern to the local human inhabitants of a region. In the case of the Mukogodo Hills, the region was historically occupied by both foragers and pastoralists. Foragers are possibly more flexible than pastoralists in that they likely adjust their dietary patterns as natural vegetation and animal regimes alter. Pastoralists, on the other hand, depend almost exclusively on their domesticated livestock, which are far less flexible than humans in their ecological requirements. Interestingly, contemporary pastoralists have shown a great deal of flexibility in dealing with localized environmental changes by altering the species composition of their herd to contain animals better suited to drier conditions (*i.e.*, goats, camels, and sheep.). In any case, it is clear that both foragers and pastoralists live their lives in close dependency upon their environment. It is reasonable, then, that the humans who have occupied the Mukogodo Hills region during the Pleistocene are better understood if we have some idea of the environmental changes that the area has undergone. The next section reviews our current understanding of paleoenvironments in the study area over the last 300,000 years, based on the types of proxy evidence discussed above.

PLEISTOCENE ENVIRONMENTS

300 Thousand Years Ago to 25,000 Years Ago (Stages 3—8)

Beginning about 300,000 years ago, a reliable high-resolution record of global ice-advance exists that is based on oxygen-isotope concentrations in deep sea sediments (Martinson et al. 1987; Pisias et al. 1984). Though shorter than the Imbrie (et al.1984) oxygen-isotope chronology, the Martinson chronology is potentially more accurate because error has been minimized by "tuning" the chronostratigraphy with the astronomical phenomenon first observed by Milankovitch (Hays et al. 1976). Furthermore, the Martinson chronology extends almost 300,000 years, which fully covers the period of interest in the study area.

The period between 300,000 years ago and the Late Pleistocene glaciation can be roughly divided into two warm and three cold periods. Each of these periods contains substantial variation and cannot be correlated to single, large-scale glacial or interstadial event. However, it is likely that these warm and cold periods were correlated with regional moisture budgets, vegetation regimes, and ice advances. In short, the oxygen-isotope record begins to illuminate an underlying ecological variable. Periods of ice accumulation that signal cool temperatures extend between 300,000 and 245,000 B.P. (event 8), from 186,000 to 128,000 B.P. (event 6), and again from 71,000 to 59,000 B.P. (event 4) (Martinson et al. 1987:19).

Radiometric dating of lake-level strand lines indicates that high water levels correspond to warmer periods when worldwide glaciers were small. Examples from East African lakes with low levels during the late Pleistocene confirm this fact. However, few long-term lake-level records exist in East Africa that extend very far into the Pleistocene. Data from Lakes Turkana, Naivasha, Elmenteita, and Nakuru present a fragmentary record back to about 90,000 B.P., but correlate well with long-term records from North Africa over the same period (Butzer et al. 1972; Cassanova and Hillaire-Marcel 1992; Owen et al. 1982; Szabo et al. 1995). Figure 4 shows episodic pluvial periods from north and East Africa over the past 300,000 years. Correlations are clearly evident between wet periods in East Africa, and warmer weather and climatic amelioration (events 1, 5, and 7). However, high lake levels exist in some north African lakes during cold periods, notably event 6, but they do not begin to rise until after peak glaciation and (presumed) ablation has begun. This led Szabo et al. (1995:239) to conclude that pluvial activity is directly correlated with the major interglacial marine oxygen isotope stages (events 1, 5, and 7). If this correlation is consistent, it may hold true that low lake levels elsewhere in East Africa are correlated to periods of ice accumulation. All studied lakes in Kenya, including Naivasha, Nakuru, Elmenteita, and Turkana, and others in East Africa, were much lower during the last ice age, lending support to this idea (Haberyan and Hecky 1987; Richardson and Richardson 1972; Street-Perrott et al. 1989). Therefore, periods of ice accumulation in East Africa are periods of increased aridity, while the converse is true during ice-free years.

Glaciations on Mt. Kenya provide a general framework for understanding long-term regional climate change in the study area. It is important to recognize that because glacial advance is a destructive process, a chronology of climate based on glacial advance is necessarily coarse, principally based on the largest glaciations that left prominent tills and moraines for later identification. The peak of Mt. Kenya lies approximately 280 km southeast of the Mukogodo Hills region. During the maximum extent of Pleistocene glaciation, glaciers descended to

2,850 m on the western flanks of Mt. Kenya (Mahaney et al. 1990:286). At the end of the Pleistocene, the maximum extent of glaciation was somewhat less at about 3,200 m. The distance from the study area to glaciated areas is therefore no more than 50 km, and less in some places Therefore, climatic reconstructions based upon Mt. Kenya glaciations are highly relevant.

Because Late Pleistocene glaciation created some very well preserved tills, it is perhaps the best understood period of Pleistocene glaciation. On Mt. Kenya, this advance is collectively known as the Liki Glaciation, but is otherwise broken into three Liki substages (Liki I, II, and III). It is likely that throughout the Pleistocene, general warm and cool periods experienced a great deal of relative temperature change. Indeed, this is supported by the oxygen-isotope record that shows minor and major fluctuations within each "event" (Martinson et al. 1987). From a larger perspective, periods covered by such long-term events as the Liki Glaciation can be considered as a single event. The earliest of these is Liki I, which is thought to have begun about 100,000 years ago (Mahaney 1990:133). Other than correspondence to the oxygen-isotope curve, it is very difficult to approximate the inception of the Liki Glaciation. The normalized $\delta 18O$ curve indicates a resurgence in cold air temperatures beginning about 74,000 years ago (Martinson et al. 1987). This is generally thought of as the beginning of the last worldwide glacial advance knows as the Wisconsin or Würm glaciation. The earliest radiocarbon age estimate of Liki Glaciation is 33,840 ± 2,810 B.P. from Liki II sediments, above thick till deposition (Mahaney 1990:133). The thick underlying till indicates that glaciation significantly precedes that age. Mahaney notes that rates of sedimentation and weathering characters suggest an early date for the onset of cooler temperatures, probably much closer to the onset of global cooling about 74,000 years ago.

Last Glacial Maximum to The Holocene: 25,000 Years to 12,500 Years Ago (Stage 2)

Radiocarbon dating and the existence of well-preserved vegetation data in the form of pollen records allow the climate of last 25,000 years to be particularly well-understood. According to pollen evidence from Kamiranzovu Swamp in Rwanda, the period of greatest coldness and aridity in East Africa dates between 21,000 B.P. and 14,500 B.P. (Hamilton 1982:191). However, it should also be noted that the pollen evidence indicates even drier climates during this period in Kenya. Though Kamiranzovu is slightly westward of the study area, its pollen record is long and compares well with other high altitude sites from East Africa such as Lake Kimilili and Koitoboss Bog of Mt. Elgon, and Sacred Lake and Lake Rutundu of Mt. Kenya (Hamilton 1982).

Pollen studies show that forests were greatly reduced during this cold period. The pollen diagram for Laboot Swamp, Mt. Elgon, at 2,880 m, shows between about 23,000 B.P. and 14,000 B.P. is conspicuously full of grasses (*Gramineae*), while montane types are notably absent (Hamilton 1982:131-134). This contrasts somewhat with Mt. Kenya, which did, in fact, show some evidence for montane vegetation types during this period.

Some researchers have described the vegetation zones that change with elevation on East African mountains as "belts." This has become a convenient way to discuss climate change because it is presumed that as temperature and precipitation varies, these vegetation belts will shift elevation. Sacred Lake, on Mt. Kenya at 2,400 m, is today located in the Montane Forest Belt. The pollen diagram for Sacred Lake shows significant amounts of arid adapted genera (*Artemisia*, *Anthospermum*, and *Gramineae*) between 23,000 and 14,000 B.P. (Coetzee 1964). Hamilton (1982) notes that the presence of *Artemisia*, *Anthospermum*, and other members of the Ericaceous Belt vegetation must indicate an on-site presence

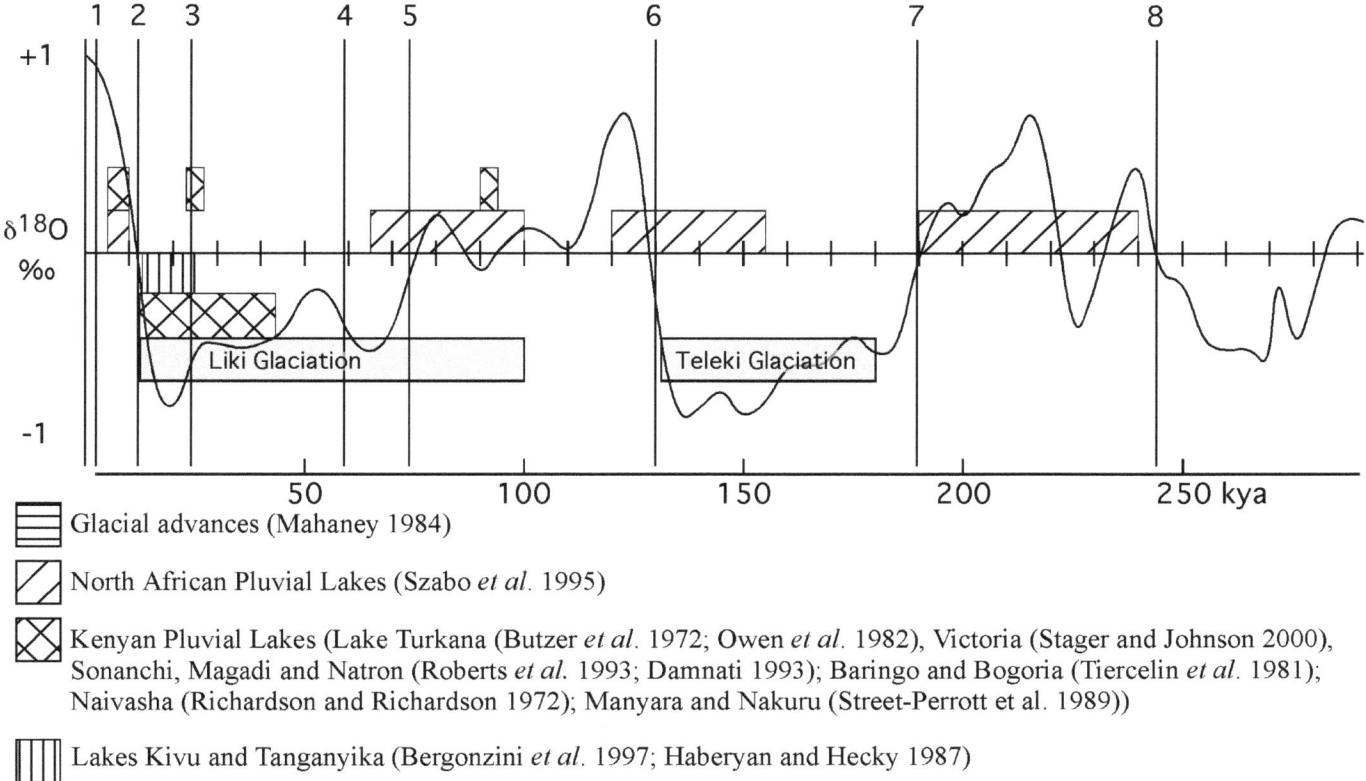

Figure 4. Oxygen isotope curve based on data in Martinsen et al. (1987). Vertical tick marks indicate 10,000 years. Major (numbered) divisions indicate isotopic stages. Isotope incursions above the line indicate periods of warming and high lake levels. Isotope incursions beneath the line indicate periods of cooling, low lake levels, and glacial advance.

of those vegetation types. If temperature change were responsible for the lowering of the Ericaceous Belt to encompass Sacred Lake, this would imply a temperature at least 4.4° C below today's temperatures (Hamilton 1982:142).

We can conclude that temperatures were substantially cooler, and P-E substantially lower, between 21,000 and 14,500 B.P. This is consistent with glacial evidence that this period coincided with the glacial maximum. On Mount Kenya, for example, the Liki II glaciation was at its maximum during this period, though having begun as early as 25,000 B.P. and having ended about 15,000 B.P. (Mahaney 1990:133-140). Similar glaciations occurred on other regional ranges including the Aberdares, Kilimanjaro, and Elgon (Hamilton 1982; Mahaney et al. 1990; Osmaston 1989).

Some East African lake levels at this time were at their lowest. Sedimentary and diatom analyses show that Lake Tanganyika was lower than present levels between 25,000 and 13,000 B.P. (Bergonzini et al. 1997:301). The shallowest conditions occurred between 21,000–16,000 B.P., with the minimum level centered at 18,000 B.P.

Pleistocene/Holocene Transition to Mid-Holocene: 12,500 to 5000 Years Ago (Stage 1)

By about 12,500 B.P., Kenyan glaciers were in full retreat and the regional climate was warming. Basal dates from bog core, and other basin substrates in Liki II end moraines on Mt. Kenya have been dated between 10,380 ± 250 B.P. and 12,590 ± 300 B.P.

Palynological evidence also indicates drying conditions at this time. The end of Liki II glaciation on Mt. Kenya coincides with reductions in more arid adapted genera (*Artemisia*, *Anthospermum*, *Ericaceae*, *Stoebe*, and *Cliffortia*), and a noted increase in humid adapted genera (*Hagenia*, *Afrocrania*, *Olea*, *Prunus*, and *Podocarpus*) (Coetzee 1964). These assemblages indicate rising temperatures, and locally, the replacement of xeric vegetation with a more mesic forest (Coetzee 1967; Lind and Morrisson 1974:195). Presumably, Ericaceous belt vegetation moved upwards on Mt. Kenya, displacing some Alpine vegetation. Lake Rutundu, which lies 740 m higher than Sacred lake on Mt. Kenya, lies fully within the modern Ericaceous Belt and does not show any signs of major vegetation changes during the Holocene (Hamilton 1982:148). It is noted, though, that long distance montane elements increase during the Holocene, confirming the changes noted at Sacred Lake.

East African lake levels provide more direct correlates of the regional moisture budget. At about 12,400 B.P., extreme aridity and evaporative conditions were present at Lake Victoria (Stager and Johnson 2000). These conditions persisted until about 10,000 B.P., indicating that even though glaciers were in retreat, the hydrological system took some time to fully adjust. Radiocarbon and uranium-thorium dating of high lake-level strands show that several high stands occurred in East African lakes between 6,000 and 10,000 B.P. (Butzer et al. 1972). Lake Turkana reached its maximum stand by 9,500 B.P. and remained high until after 7,500 B.P., with a brief transgression again to high levels between 6,500 and 6,200 B.P. (Butzer et al. 1972). From before 9,200 B.P. until about 5,650 B.P., Lake Naivasha was significantly larger than it is at present (Richardson and Richardson 1972). Richardson and Richardson suggested that high lake levels at Naivasha would have required an increase in rainfall equal to about 65% above the modern average. High level stands have been documented at some of the other lakes in the Eastern Rift, including Lakes Nakuru, Elmenteita, and Magadi (Butzer et al. 1972; Washbourn 1967; Washbourn-Kamau 1970, 1975). Other evidence, such as diatomic species concentrations, further suggest that P-E increased into the mid-Holocene, around 5,000 B.P. (Stager et al. 1997; Stager and Johnson 2000; Stager et al. 1986). Thus, glacial, pollen, and lake-level data can be used to infer that at the beginning of the Holocene, conditions in East Africa became warmer and wetter.

Mid-Holocene to Present:
5000 B.P. to Present (Stage 1 Continued)

Humidity peaks during the mid-Holocene. The early Holocene was characterized by mass wasting, but beginning about 5,000 years ago, the slopes begin to stabilize (Mahaney 1990:289).

A pollen diagram from Hobley Valley on Mt. Kenya shows declining levels of *Podocarpus* until after 3,500 B.P., then a sharp increase in *Podocarpus*, accompanied by a rise in *Myrica* and *Olea*, and a decline in *Gramineae* (Mahaney 1990:201). Taken together, this change in assemblages indicates a moist vegetation regime being replaced by a dryer one beginning about 3,500 B.P. Evidence from Sacred Lake also suggests a climax mesic forest between 5,000 and 2,000 B.P. (Coetzee 1964). This reveals a later change than other evidence indicates, but this delay in vegetation shift may be partly due to the high elevation of the Sacred Lake (2,200 m) and even higher Hobley Valley bog core (4,090 m).

It is during this period that humans in East Africa probably began producing rather than collecting food. Both pastoralism and horticulture are believed to have risen in East Africa during this time (Harlan 1992; Marshall 1994). It is important, therefore, to evaluate the possible effect that these cultural practices may have had on the landscape. Pollen evidence must be reconciled with other paleoenvironmental indicators.

At Lake Turkana, for example, a combination of pollen evidence and dated strand-lines has demonstrated that although humidity was high at the beginning of the Holocene, by the mid to late Holocene the vegetation had reached modern arid proportions (Owen et al. 1982). Especially significant are the expansion of grasses during the mid-Holocene, a necessary precursor for the early introduction of cattle pastoralism in that area (Marshall et al. 1984; Owen et al. 1982).

COLONIAL-ERA ECOLOGICAL TRENDS

Until recently, the main sources of Colonial-era ecological trends were the diaries and other publications by early settlers and explorers. McGregor Ross was a British assistant engineer of the Mombassa-Uganda railway construction who continued to be a public servant in the East African Colony for twenty-three years. He writes a generalized description of Laikipia that is recognizable to the modern visitor:

> Immediately after rains, light though they are, the "Taru Desert" in this belt is an expanse of green bushes interspersed with masses of white and purple convolvulous and other flowers. The leaves and flowers soon vanish, however, and for a period of months the deeply rooted bushes fight a battle for life with blazing sunshine and high winds. As the 2,000-foot level is reached the heat is palpably less continuous. Evenings are chill and dews frequent. Wilderness conditions disappear. Scattered mountain masses throw down some small streams. The low scrub jungle is never entirely leafless, and open patches of grassland and turf appear [Ross 1927:28].

Such descriptions are interesting, but provide little insight into the ecological conditions that prevailed during colonial times. Sir Charles Eliot, His Majesty's Commissioner of the East Africa Protectorate between 1901 and 1904, relates a slightly more detailed account:

> North of Mzara the country consist of plains which are in parts grassy, but often strewn with boulders of lava or granite. Game, however, is abundant. To the west are several mountains, of which the best known is Donyo Girri-Girri, near Laikipia. The most important physical feature in the whole district is, however, the Gwaso-Nyiro, which receives the Gwaso-Narok from Laikipia, and the Gwaso-Marra from the direction of Kenya...To the north of Gwaso Nyiro lies an undulating desert, covered with granite boulders and thorn bushes. There is no water in the rivers during the dry season, but it can be obtained by digging [Eliot 1905:75-76].

The "Gwaso-Nyiro" referred to in the above passage is the Ewaso Ng'iro of today's usage. The thorn bushes of the Ewaso Ng'iro Plain are possibly varieties of *Acacia* similar to those found there today. There is no mention of tall grasses in that region which is at odds with what some local informants have suggested. The descriptions above lead one to believe that the "desert" described has not changed much since Colonial times. Indeed, the British were more interested in the more agriculturally productive areas.

One area subjected to the most intensive colonial impact was the highland farmland of the Kikuyu. It has been argued that indigenous agroforestry practices of the Kikuyu had, prior to colonial intervention, mitigated the impact of deforestation by incorporating multipurpose trees into the local system of production (Castro 1993:45). Colonial agricultural practices were often at odds with these conservatory practices. Europeans imported coffee, wheat, cattle, and a variety of other commodities for production with varying degrees of success. Consequently large expanses of land were put under the plow or otherwise cleared for grazing. Displaced indigenous agriculturalist and pastoralist peoples, being restricted to certain areas, were forced to more intensively utilize the landscape in areas that they were restricted to. After independence was achieved in 1962, some farmers were able to return to their traditional lands, but in most cases continued the tradition established by the colonial administrators. In some areas it is still illegal to destroy a coffee plant, though many farmers get around this legislation by seasonally pruning them to the ground. It has been shown that between 1986 and 1992, the woody plant (tree) biomass in agricultural areas has increased (Holmgren et al. 1994:390). This growth is attributed to indigenous land-tenure systems, indicating that Kenyan farmers do indeed practice sustainable management. Some areas considered poorly suited for agriculture in the 1930s are today highly productive agricultural centers. Machakos District is a semiarid region in central Kenya. Due to the introduction of drought tolerant species, irrigation, and intensive labor, this area today supports five times the population of 50 years ago, and agricultural production is estimated to be as much as fifteen times higher (English et al. 1994).

The well-watered upland areas of central Kenya, which include the headwaters region of the Seaku River in the Mukogodo Hills, are excellent pastures. Though this was an area that the Maasai were originally guaranteed perpetual grazing access under a treaty with the British, it was later reacquired by the Colonial government for agricultural purposes (Ward and White 1971:107). In areas such as these agriculturalists can come into conflict with wildlife. Poaching of large game is considered a major problem by the Kenya Wildlife Service, but to ranchers, pastoralists, and small-scale farmers, big game can be considered a major pest. There is little historical documentation of this kind of conflict other than passive mention, but the ubiquity of recent accounts does not lead one to conclude that this is solely a modern phenomenon. Today, approximately 3000 elephants live in the Laikipia-Samburu region of northern Kenya, which includes our study area (Thouless 1994:119). These elephants are known to cause a variety of problems for the people living in the area. As such, each year a number of elephants are shot, poisoned, snared, or otherwise eliminated by the local people (Thouless 1994:119). The assault on elephants, rhinos, and other game animals both directly and by ecological disturbance has caused their numbers to dwindle.

An elderly Samburu woman living in Kipsing remembers the Ewaso Ng'iro Plains as it was 40 to 50 years ago: "It was not thick bush like it is today," she said. "Then it was just young acacia and open grassland. It was the home of elephants and rhinos, but today there are only a few elephants that come to drink by the Seaku. In those days, after the long rains the grass grew to be very tall. When walking around the tall grass could reach your eyes! There was plenty of grass" (M. Lemungesi, translated taped interview, August 2000).

This statement of her childhood memories may be nostalgic. In the absence of environmental data, her statements would be difficult to prove. They are, however, consistent with expected changes that would result from declining populations of megaherbivores and increasing populations of domestic livestock. The extent to which pastoralist populations have grown in the last 50 years due to the unintended consequences of colonial era government policies is dramatic (Cronk 2000). If humans have impacted the environment, then evidence might be preserved in the geological record of erosion, deposition, and stability of the landscape. This subject will be considered in the following chapter.

CHAPTER 3
GEOARCHAEOLOGY

Geoarchaeology is the application of geological studies to the interpretation of the archaeological record. Geoarchaeological investigations have been an important component of the Mukogodo Hills archaeological research project since its inception. Initially, emphasis was placed on interpreting stratified rockshelter sediments from Shurmai Rockshelter (Kuehn and Dickson 1999). More recently, the focus has shifted to the alluvial sediments extending from the Mukogodo foothills, where Shurmai is located, north to the Kipsing River about 10 km away. Current geoarchaeological investigations in the project area have:
- Establish an alluvial chronology to assist in temporally placing archaeological sites discovered in buried contexts.
- Determine to what extent archaeological site distribution and preservation has been affected by ongoing natural site formation processes.

This work began in 1995 with the recording of a single large section of the Tol River by Dave Kuehn and Steve Jennings (Kuehn et al. 1996). In 1996, Kuehn recorded several other geologic sections along the Tol in an effort to better understand the alluvial chronology of the Seaku drainage system, and by reference to the wider Kipsing River basin. I joined the project in 1996, and with Kuehn, recorded some of the natural exposures in the field. In 1999 and 2000, I returned to the field collecting additional geological data and field checking previous results.

In total, eleven natural cutbank exposures have been examined and profiled in our study area (Figure 5). Each of these were cleaned by hand, then profiled and photographed. The soil descriptions conform to standards set forth in *Keys to Soil Taxonomy* (SCS 1992), while non-soil stratigraphy was described using the North American Stratigraphic Code as a guideline. Soil horizons and lithologic units were characterized based on their lithology, color, texture, structure, and other soil characteristics. Particular attention was paid to secondary carbonate formation. In a few cases, additional granulometric and chemical analyses were undertaken. When possible, samples for radiocarbon dating were collected. Detailed descriptions of the profiles was reported in Pearl (2001), appendix A. What follows here is a more detailed discussion and interpretation of the geoarchaeological investigations.

THE 1995 TOL RIVER PROFILE

The central tributary in the study area is the Tol River, a 20 km long, third order stream that forms a part of the Ewaso Ng'iro River system of central Kenya. Tol River stratigraphy shows a complex picture of deposition, stability, and erosion. Kuehn et al. (1996) identified 10 lithostratigraphic units in the 4 m Tol River profile excavated in 1995 (Figure 6). Additionally, six soils were identified, each representing a period of surface stability, though each does not represent a paleo-surface as some soils have been truncated by erosion.

Three radiocarbon dates provided an indication of the time depth expressed in the exposure. A radiocarbon sample of organic carbon obtained from a sample of indurated carbonate nodules at the base of the profile yielded an age of 19,940 ± 140 B.P. (CAMS-33987). Bulk soil organic matter 60 cm higher in the profile yielded a radiocarbon age estimate of 8470 ± 350 B.P. (GX-21706), revealing that most of the exposure could be assigned to the Holocene. Radiocarbon dates on charcoal of 1390 ± 95 B.P. (AA-20481) and 405 ± 90 B.P. (Lab no. unknown) within approximately 2 m of the surface demonstrate that much of the exposure dates, in fact, from the mid to late Holocene. Radiocarbon results are provided in Table 1.

Kuehn et al. (1996) hypothesized that the profile showed three important events: 1) extreme aridity during the late Pleistocene, around 20,000 years ago; 2) expansion of grasslands about 8,500 years ago; and 3) the subsequent introduction of pastoralism and widespread exploitation of the grassland environment, expressed as erosion in the alluvial stratigraphy. However, there is a problem with this interpretation. If the upper 2.5 m is less than about 8,500 years old, and the top 1.0 meter is less than 500 years old, then only late Iron Age artifacts should be found on the surface. Site survey in 1996 revealed that a broad range of artifacts occurred on the surface, including artifacts that were morphologically similar to Middle and Later Stone Age artifacts recovered from Shurmai and Kakwa Lelash rockshelters (Gang 1997). Questions concerning the lateral extent of the upper member of the Tol River arose, and beginning in 1999, an effort was made to correlate surface finds with the preliminary alluvial chronology outlined in the Tol River section.

IMPORTANT GEOLOGICAL PROCESSES IN THE MUKOGODO FOOTHILLS AND EWASO NG'IRO PLAINS

Fluvial processes are the most basic geomorphic processes observed in the study area. Fluvial processes are those natural land-forming processes directly associated with the flow of water. Fluvial landforms in the study area include channels, terraces, fans, and pediments. In geoarchaeological terms, each of these landform types can be seen as a different natural site formation environment, where site preservation characteristics are highly varied. One of the goals of geoarchaeology is to understand the impact of these varied processes on the archaeological record. In fact, this is an absolutely mandatory practice if any sense is to be made of site distribution in an arid environment.

In our study area sites of different ages (i.e. Middle and Later Stone Age, etc.) are unevenly distributed. Before considering a cultural explanation, we must first evaluate the possibility that the site distribution is due to natural site formation processes. Studies in the arid southwestern United States have demonstrated the significance of this approach. For example, research has shown that variability

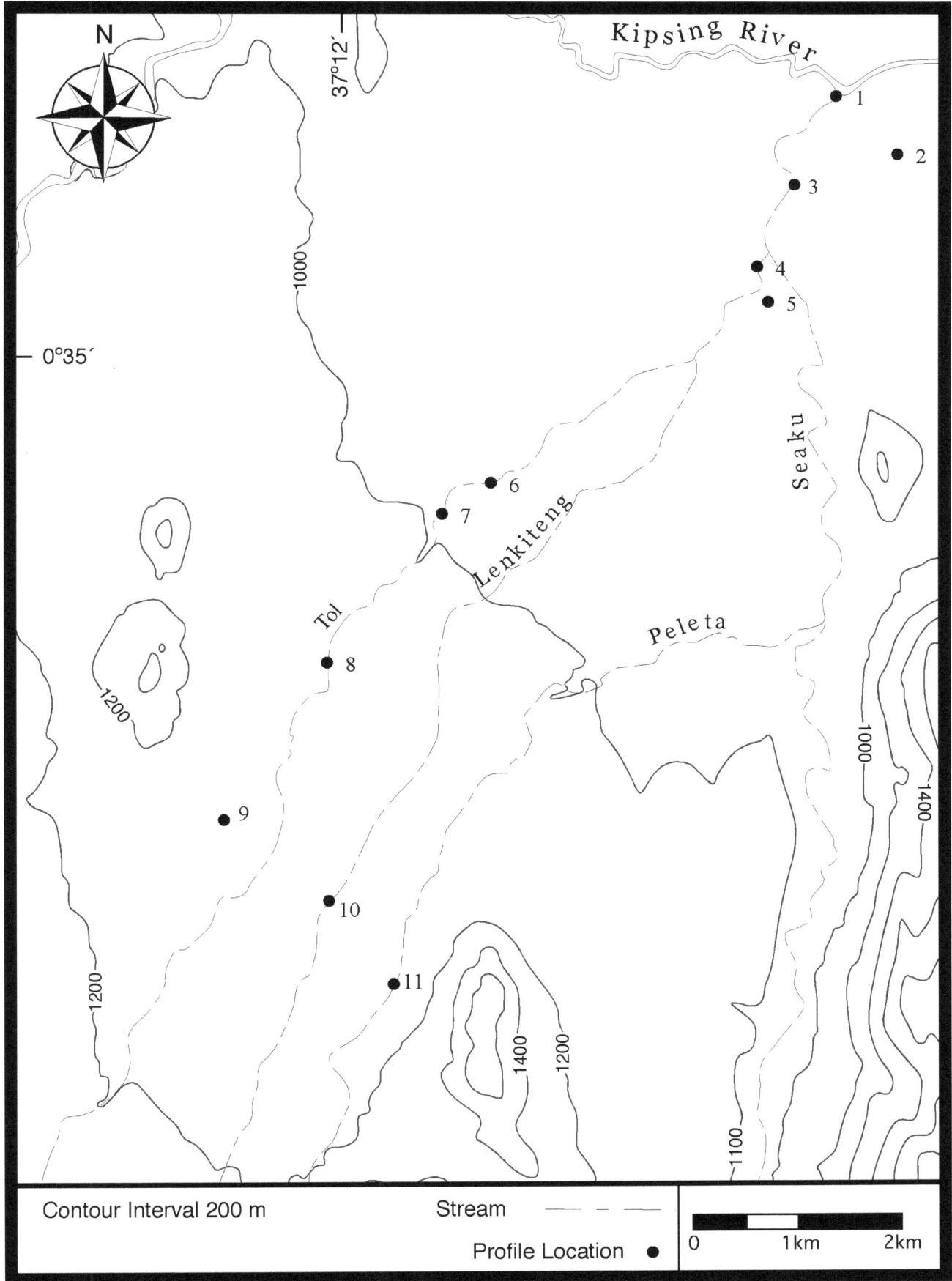

Figure 5. Locations of sedimentary profiles described in the text.

Table 1: Radiocarbon Results

Context	Age (BP)	δ¹³C_{PDB}(‰)	C-13 Corrected	Lab Number	Material Tested
Qt	116.5±0.7		yes	A-12000	charcoal
Qt	405±90	-	-	unknown	dispersed charcoal
Qsk	1,390±95	-23.2	yes	AA-20481	charcoal
Qsh	8,470±350	-13.9	yes	GX-21706	soil humates
Qp	19,940±140	-15.6	yes	CAMS-33987	organic residue
Qp	37,710±4200-2740	-0.7	yes	GX-21652-C	carbonates

in Clovis culture sites was purely a product of these natural site formation processes and not the result of cultural land-use preferences (Waters 2000; Waters and Kuehn 1996).

Fluvial Channel and Terrace Sediments

Any point on any landscape exists in one of three states: deposition, erosion, or stability. A drainage basin is an extremely dynamic place. Sediments in the stream must have come from somewhere - either from unconsolidated bedload upstream, or from direction erosion of consolidated terraces, or from bedrock upstream. Fluvial sedimentary transport occurs when the stream power (usually characterized by its velocity) exceeds the cohesiveness and/or mass of the sediments. Sediments are deposited in channels or floodplains, or are eventually transported downstream. As velocity decreases, the heaviest particles are the first to drop out of entrainment (motion). This results in a fining upward sequence of sediments as the velocity of a stream decreases after a flood. At times, stability may dominate a region allowing vegetation growth and soils development. The combination of these processes over time results in terraces with distinct textural compositions, often with soils buried within them. These stratigraphic subdivisions might be traced from terrace to terrace, and even from stream to stream.

Deposition and erosion are interrelated responses that occur within a dynamic system. In arid lands with light vegetation cover, sporadic rainfall and flashy discharge can cause erosion and downcutting into terrace sediments. Conversely, low-level flow with regular flooding can result in deposition and accumulation of sediments. The history of deposition, stability, and erosion are recorded in the soils and sediments that underlie fluvial terraces. Terrace remnants might be seen along valley walls or preserved in subtle step-like arrangements on the terrace surfaces themselves. Terrace levels are often consistent through a drainage network, and if they can be dated, a chronological framework outlining the ages of the terraces can be developed. Holding other factors such as bedrock, climate, and vegetation constant, there might be a great deal of regional correlation in a drainage network.

Coarse stream deposits indicate high velocity deposition. These are most often channel deposits, such as bars. In the contemporary setting, these resource areas are actively used by local pastoralists, principally for water procurement but also for other activities as well. Unfortunately, archaeological sites created in these areas are usually destroyed by subsequent high velocity flow. Furthermore, sites are often eroded from upstream locations and redeposited in the channel sediments. Therefore, sites discovered in coarse channel sediments are likely to be in a secondary context, possibly having been transported very far from their initial location of deposition. Furthermore, due to sorting by fluvial processes, such secondary sites have been severely altered. In contrast, sites buried by low velocity fluvial processes are typically better preserved, having been subjected to fewer of the impacts associated with high energy processes. An archaeologist must carefully consider the possible effects of fluvial transport on transforming archaeological sites. All fluvial sediments are good candidates for site burial.

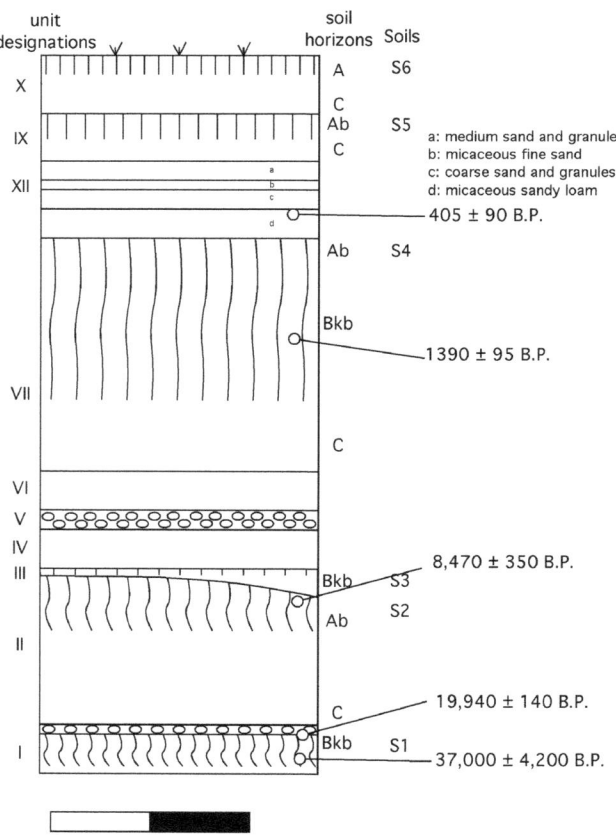

Figure 6. drawing of the Tol River profile based on Kuehn's 1995 notes.

Alluvial Fans

An alluvial fan is a sloping mass of alluvium that was deposited as a result of flow from a high gradient tributary to a lower gradient one, or onto some other geologic "floor." In the study area, high gradient streams flow down from the Mukogodo Hills and onto the Ewaso Ng'iro Plains. It is precisely at this juncture, where the streams are changing gradients, that alluvial fans have formed. These deposits are often referred to as colluvium or "slope wash." The fans themselves are very large features, sometimes extending several miles from the mountains. Exceptionally large fans are often recognizable only from the air or from a distant vantage point. The current study area can be thought of as a large alluvial fan crosscut by individual channels that have characteristics of alluvium (as described above). Thus, the term "fan" is used here to describe visibly sloping deposits at the valley margins.

Technically, alluvial fans are fluvial landforms. That is, they are formed principally through the transport of sediments by water. Due to the slope of alluvial fans and the velocity with which water sometimes enters the fan environment, fan sediments tend to be coarser. Also, mass wasting (landslides) can be a significant formation process. Fans observed in the Mukogodo Hills are generally derived from ephemeral stream flow, also known as "dry fans" (Schumm 1977). Active alluvial fans are cumulic, or constantly forming over a long period of time. In the Mukogodo Hills area, dry fans are also well drained and potentially very good locations for archaeological sites. Fan deposits generally interfinger with adjacent deposits if those deposits are also active, often resulting in a deep and well-stratified subsurface profile. Fans which are adjacent to important resource areas should be considered as candidate locations for archaeological sites; however, archaeological interpretation of any such sites must always consider that fans are constantly forming and reforming.

In some locales in the study area, colluvium is observed along the edges of inselbergs and mountains, where it is undergoing current erosion by fluvial and aeolian processes. These preserved deposits are called "pediments" and were initially formed by the processes similar to those described above. Archaeological sites are frequently buried in these pediments.

A Note on Aeolian Processes

The most common aeolian landform is the sand dune. Due to significant devegetation in the study area, we are seeing aeolian processes rework exposed sediments, but no major aeolian landforms have been identified. In many places, vegetation cover is less than 20 percent, but in others it is as high as 80 percent. Most sediments exposed at the surface are predominately fine to medium sands, with aeolian ripple bedforms common on sandy surfaces. Wind routinely whips up small dust devils, but fluvial overland flow is still the predominant geomorphic factor in the area. Any aeolian sediments noticed on the surface were recorded as a "mantle" in profile drawings.

One concern is that "ventifacts" might have been created on the surface due to long term exposure to winds. Ventifacts are stones that have been buffeted by sands and winds to such an extent that appear to have been worked by human hands. Such stones have been found in the deserts of southern California, but none were observed in the current project area, another a sign that aeolian processes have not been significant here in the past.

Human and Animal Impact

Humans and animals, especially herbivores, have a tremendous impact on the environment. Large herbivores such as elephants help maintain a diverse savanna habitat. Foragers can affect their physical environment by exploiting or overexploiting certain species, or even more directly through actions such as intentionally burning woody habitats to increase savanna productivity. Pastoralist activities affect the environment even more profoundly due to the introduction of large numbers of domesticated herbivores (see p. 26).

Measuring the impact of cattle grazing on the environment in contemporary populations is no easy task. From an archaeological or geological perspective, measuring their impact on the environment is certainly more complicated. It is thus reasonable to expect that the introduction of pastoralism in prehistory changed the environment in our study area in some ways. If that initial impact was slight, we might not see any evidence of the introduction. However, if the environment was appreciably stressed by the introduction of cattle, we might be able to find some evidence in the late Holocene geologic record.

KIPSING AREA QUATERNARY GEOLOGY

The basement rock beneath Quaternary sediments in the project area are of Precambrian age and consist largely of metamorphic rocks, especially gneisses and migmatites (Hackman et al. 1989). In the Mukogodo Hills area, these rocks are principally quartzo-feldspathic and banded biotite gneisses. These gneisses constitute the source of the Quaternary fluvial sediments discussed below. Although it is unclear precisely how deep the basement is in the Kipsing/Mukogodo Hills area, it is clearly shallow at many points, and is perhaps shallow everywhere throughout the study area. Inselbergs, or isolated granitic hills, dot the surface throughout the area. Bedrock partly determines stream gradients as the basin descends from the uplands of the Mukogodo Forest to the Ewaso Ng'iro Plains. Exposed bedrock surfaces are ubiquitous there. In some places, the bedrock is marked by nickpoints through which fluvial channels must pass and behind which groundwater accumulates.

"Alluvium" is a general term that refers to sediments deposited on floodplains, channels, and on alluvial fans. Although I have grouped together deposits of the same age that were deposited by the same process, there is some variability within these deposits. For example, an alluvial group might have a fine-grained floodplain component and a coarse channel component. Both are approximately the same age and were deposited through the same process. These different components of the same alluvial unit are called facies. The major alluvium groups were given names as a convenient mnemonic device to aid discussion. Use of names instead of numbers for alluvial units makes renaming the units easier if new stratigraphic units are found that must be inserted into the sequence. Name selection reflects the character of the local area, usually coinciding with a major topographic feature in the same vicinity.

Peleta Group Alluvium (Qp and Qpf)

The Peleta alluvium consists of a floodplain and channel facies (Qp) and a fan facies (Qpf). The Peleta group has alluvial fan, channel, and floodplain facies preserved in different parts of the project area. Its contemporary surface expression is predominately away from the major streams, close to the foothills of the Mukogodo Hills. It is also present in preserved fan deposits around some of the larger inselbergs. In the past, however, this distinctive alluvium filled much of the valley. Sediment textures range from poorly sorted gravelly sandy loams to medium loams that range in color from deep red to yellowish-red.

At one time a thick soil (S1) formed on the surface with substantial subsurface accumulation of calcium carbonate ($CaCO_3$). This buried layer of carbonates is called a calcic horizon, and is typical of aridisols, or soils formed in arid moisture regimes. However, this soil developed only weakly on the alluvial fan facies. Subsequent gullies and streams cut deeply into these deposits or buried them, but some are now exposed in the bottom of natural exposures (profiles 4, 6, 7, 8, and 9).

The Tol River profile was revisited and reexamined in 2000 (Figure 7). This exposure (profile 8) provides the best example of the Peleta paleosol, and also demonstrates its relationship to the overlying units. Here, the Peleta paleosol is described as a reddish-yellow (5YR 6/6) loam with abundant $CaCO_3$ nodules up to 5 cm in diameter. Two radiocarbon dates were obtained from this unit, with the more recent of the two being 19,940 ± 140 B.P. It was obtained by dating an extract of the organic fraction in an indurated carbonate nodule. Another age estimate of 37,710 + 4200 /-2740 B.P. (GX-21652-C) was obtained by directly dating the total carbon contained in the carbonates. This gives us a general age estimate for the formation of the preserved calcic horizon in the Peleta paleosol. This great age range reflects two things: 1) the gross nature of radiocarbon dating of carbonates; and 2) the fact that the event being dated is diachronic, that is, a long term event – the soil formed over a long period of time. Formation of the Peleta paleosol was most likely coincident with the Liki glaciation recorded in the geology of nearby Mt. Kenya.

The aforementioned radiocarbon dates that give the lower bracket of ages for this unit derived directly from the calcic horizon. This stratum is an easily recognizable marker horizon that can be traced throughout the project area. It was found 3.45 m below surface in profile 6, 3.92 m below surface in profile 7, and 1.15 m below surface in profile 9.

The Peleta fan facies (Qpf) is found at the surface in several places in the project area. Though it has not yet been mapped, it generally occurs adjacent to the Mukogodo Hills. It lacks the characteristic calcic soil horizon of the floodplain facies, yet it retains a distinctively reddish hue. Where Peleta alluvial fan facies are exposed on the modern surface, they are called the "Peleta Surface." This surface is shown in profiles 2 and 11. The weathered granitic source material is near to fans, and the surface gradient there can be substantial (3 percent at profile 2, and 9 percent at profile 11). As a consequence, sediment textures tend to be coarser throughout the facies.

Kipsing Alluvium (Qk and Qkc)

Peleta alluvium is overlain by Kipsing alluvium. The Kipsing alluvium is a grayish-white to pale brown sandy loam (Qk). It obtains its pale color from the physical weathering of surficial and buried calcium deposits. The age or the source material for the Kipsing unit is unknown, but it is most likely a fluvial deposit brought in by an ancient Kipsing River. It is calcium rich, indurated calcium concrete, otherwise known as calcrete, lies beneath its surface in some locations.

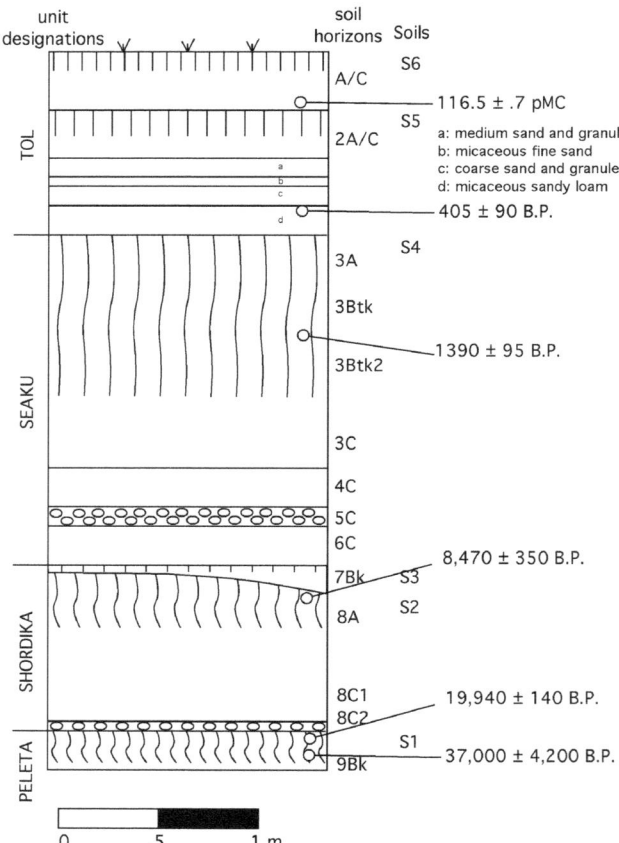

Figure 7. Tol River profile based on 2000 interpretation.

Kipsing calcrete (Qkc) does not usually form at the surface. Rather, it is a cemented conglomerate of sediment and precipitated calcium carbonate that forms beneath the surface. Interestingly, the calcrete tends to be at a relatively constant elevation. It is found at the surface near the Kipsing River, but is buried about 10 m deep only 3.5 km upstream in the Tol River at approximately the same elevation. This is consistent with the change in surface elevation between the two areas. The soil above the calcrete is a mixture of weathered calcrete residuum and the original fluvial deposits. In some places much of the original surface has been eroded down to the resistant calcrete.

Peleta alluvium is not found throughout the project area, but only near the Kipsing River. Along the lower Seaku River near its junction with the Kipsing River, there are good exposures of Kipsing alluvium (profiles 3 and 4). It is characterized as a very pale brown fine sandy loam. The calcrete appears 1.5 m at greater depth and is the same color.

The presence of calcrete deposits presents an interpretive problem because no one knows with certainty how long it takes calcrete to form. It takes a minimum of 10,000 years for mature calcrete deposits to form, but this should be considered a conservative estimate for thick deposits. (Gile et al. 1966). Although the true thickness is unknown, blocks at least 2 m thick are found in the Tol River bottom. It is about 4 m thick at Kipsing. The material is highly erodable; the Tol River has easily cut through it, and in most cases it is buried with alluvium. At any rate, the surface is quite old and has the potential for very old archaeological sites within it and on its surface.

Cobbles within the calcrete indicate that at one time stream power was significant (Profile 4). Further evidence of the former stream power is seen in the smoothed granite surfaces downstream. The cobbles are basalt – the same type of material that over 90 percent of the lithic artifacts in the region are made from. This particular location does not appear to be a quarry and no other exposure has yet been discovered with basalt cobbles.

Shordika Alluvium (Qsh)

Shordika alluvium, which is principally fan deposit, is most pronounced in the Tol River uplands, well away from the Kipsing River. It is coarser and less well sorted than some of the overlying fluvial units. It ranges from reddish-brown to yellowish-brown, medium to coarse sandy loam, though some variation exists. It often appears to have a reddish caste, partly the result of in situ oxidation of mineral elements in the soil, and partly inherited from the parent material for this unit – perhaps reworked Peleta alluvium and eroded granitic gneiss. The Shordika alluvium is seen in Profiles 8 and 9.

Though the surface expression in the upper Tol and Seaku interfluvial plain is extensive, the Shordika alluvium was only captured at the surface in one exposure (Profile 9). There it was shown to directly overlie Peleta alluvium. Based on similar soil characteristics and stratigraphic position, the Shordika alluvium can be traced downstream to Profile 8 as well. A radiocarbon date on the upper portion of the Shordika alluvium was 8,470 ± 350 B.P. This date comes from a buried soil (S3), called the Shordika paleosol, that caps the unit. Neither the Shordika soil expressed at the surface in Profile 9, nor the Shordika paleosol expressed in Profile 8, are substantially developed.

Seaku Alluvium (Qsk)

The Seaku alluvium, which only occurs in the central and upper portions of the Tol River, is capped by a well-developed soil (S4). The strongest expression of this soil is seen in Profile 8, but it is also visible in Profiles 6 and 10. The 1 m thick upper portion of this buried soil, the epipedon, has a total organic carbon content up to 0.5 percent. This, among other soil factors, qualifies it as an ochric epipedon, but not a mollic epipedon (SCS 1992). A mollic epipedon requires at least 0.6 percent total organic carbon. Mollic epipedons form over great periods of time, usually under grasslands. This difference is significant because it implies that this soil was progressing to becoming a mollisol. The difference probably involved temperature, time, and water, possibly indicating an arid grassland/savanna environment. Mollisols tend to form during cool regimes, which is another reason that formation was hindered. Radiocarbon determinations indicate that the Seaku Alluvium formed after 8,470 ± 350 B.P., and before 405 ± 90 B.P. (see below). A charcoal radiocarbon age of 1,390 ± 95 B.P. (AA-20481) was obtained from the middle of the Seaku paleosol.

Although is subtle, there is some variation in soil development within the epipedon of the soil at Profile 8, indicating that there are actually two soils superimposed (welded) upon each other. The age of 1,390 ± 95 may represent a mean residency time for that soil, in which case the surface, which was stable until about 405 ± 90 B.P., was undoubtedly established earlier than 2000 B.P. Based on the level of soil development, the Seaku surface was probably stable for a long period of time.

Tol Alluvium (Qt)

The Seaku Alluvium is overlain in the Kipsing drainage by the Tol alluvium. A charcoal radiocarbon age estimate of 405 ± 90 B.P. taken from the base of the Tol unit at profile 8 indicates that these sediments were deposited over roughly the last few hundred years. A modern radiocarbon result on charcoal (A-12000, 116.5 ± 0.7 pMC) revealed that the Tol unit is still actively accreting. Tol alluvium is most common along the channels of the lower Tol and Seaku rivers where it retains visible sedimentary structures (bedding and lamination) near the surface. Outcrops of Tol alluvium occur in Profiles 1, 5, 6, 7, 8, and 10.

A radiocarbon age of 405 ± 90 B.P. from the base of the Tol unit at Profile 8 is evidence of its recent age. Nowhere is it found underlying any other unit, indicating that it is indeed the most recent stratigraphic unit in the project area. However, as aeolian deposition increases in thickness, what is now a thin "mantle" could eventually become a significant depositional unit (see profile 10).

The Tol alluvium is a brown sandy sediment dissected by shallow channels. Soil development tends to be weak or even nonexistent in some locations. The alluvium's surface expression east of the Tol River is somewhat heterogeneous as sands and muds are sorted into a microcosm of channels and pools that formed in the loosely consolidated floodplain. Indeed, it appears that the present lack of vegetation is contributing to the erosion of these floodplain sediments, which have taken on the character of a miniature braided stream environment. Vegetation is patchy but rarely exceeds about 60 percent cover, usually below 50 percent, and in some places is totally absent. West of the Tol River, the sediments are more stable where they occur, perhaps because they are interfingered with colluvium.

SUMMARY OF ALLUVIAL CHRONOLOGY

Figure 8 shows a longitudinal correlation of selected profiles in the study area. While the horizontal alignment is not to scale, the relationship between the profiles demonstrates the elevation change between profile locations. There is a general progression from very little Tol (Qt) Alluvium in the upper reaches, (profile 11) to where it dominates the profile at the junction with the Kipsing River (profile 1). Also, the Pleistocene alluvia, (Qp and Qk) have irregularly undulating surfaces. Erosion of these alluvia has been irregular, and as we shall see, variations in the underlying bedrock and indurated elements within the alluvia have affected the character of their preservation. Finally, it should be noted that no single profile shows the complete alluvial chronology. In order to understand the complex history of erosion, deposition, and stability, it is necessary to generalize the stratigraphy by making correlations between laterally distinct profiles.

Based on an expanded understanding of the regional correlation of stratigraphic units, and from the radiocarbon ages obtained in Profile 8, the alluvial chronology can now be summarized. The oldest alluvial unit in the project area is the Peleta alluvium (Figure 7). It was deposited sometime during the Quaternary in both floodplain and fan settings. During a period of stability, a thick soil formed on the level floodplain surface and weaker soil development occurred on the steeper fan deposits. The Peleta soil, an aridisol, developed over a considerable period of time under arid conditions as indicated by the well-developed calcic "carbonate" horizon. Soil formation was most likely coincident with the Liki glaciation recorded in the geology of nearby Mt. Kenya (Figure 4). This glacial event spanned a long period of time between about 13,000 and 75,000 B.P. when the regional moisture budget was locked up in glaciers on the East African mountains (Mahaney et al. 1989). The Kipsing calcrete probably formed during this period of peak aridity. The Kipsing alluvium is approximately the same age as the Peleta alluvium, or is slightly younger (the Kipsing alluvium is not shown on Figure 7 because it was absent from Profile 8).

At the end of the Pleistocene, conditions became wetter and warmer. A period of channel incision occurred in the central portions of the river valleys, cutting deeply into Peleta alluvium. Erosion resulted in the loss of Peleta alluvium along the valley centers. In many cases, erosion was hindered by the calcic horizon which was more resistant than the overlying A and B horizons. Deposition ensued, reburying the relict Peleta soil under the Shordika alluvium. When the Shordika unit stabilized, a soil formed on its surface that has been dated to $8,470 \pm 350$ B.P. It aggraded between the end of Liki glaciation until the early Holocene age as indicated by radiocarbon dating. The period of time represented by the Shordika alluvium is between about 13,000 B.P. and 8,500 B.P.

Like the Peleta surface, along the valley margins the Shordika surface is preserved in fan deposits. These conditions result in weaker soil development, and nowhere are Shordika soils substantially developed. Thus, at present, no firm conclusions can be reached regarding environmental conditions in the project area during the early Holocene.

Sometime during the middle Holocene, another major episode of floodplain aggradation occurs and the Seaku alluvium is deposited. A radiocarbon age from the center of this unit indicates that deposition was well underway by the about 1,400 B.P. A thick soil formed on its surface. It is significant that this soil that was evolving towards becoming a mollisol. Mollisols typically form in temperate grassland/savanna environments; however, a full mollisol did not form. This may indicate that conditions were too dry, too warm, or the soil simply did not have enough time to develop. In any case, large amounts of organic carbon were entering the soil column during this period.

The Tol alluvium indicates that within the last 500 years, deposition resumed in the study area. No archaeological sites have been found buried in the Tol alluvium or even upon its surface. Though conceivably such sites may exist, Iron Age and other recent sites tend to be found on older stabilized landforms. Tol alluvium represents places likely to undergo modern flooding. As such, its surface may have been avoided by late Holocene inhabitants. Alternately, because of its limited expression, we may simply not have yet found many of these sites. The upper member of Profile 8 is the toe slope of a small alluvial fan originating at Kakwa Lelash. The complexity exhibited in the Tol alluvium of Profile 8 is due to active fan deposition interfingering with occasional floodplain deposition.

PREDICTIVE SITE MODEL

The first goal of geoarchaeological investigations in the study area was to establish an alluvial chronology to assist in temporally placing archaeological sites. To that end, a predictive site model has been developed and is presented in Table 2. This model not only predicts where sites of given ages might be discovered, it also can be used to

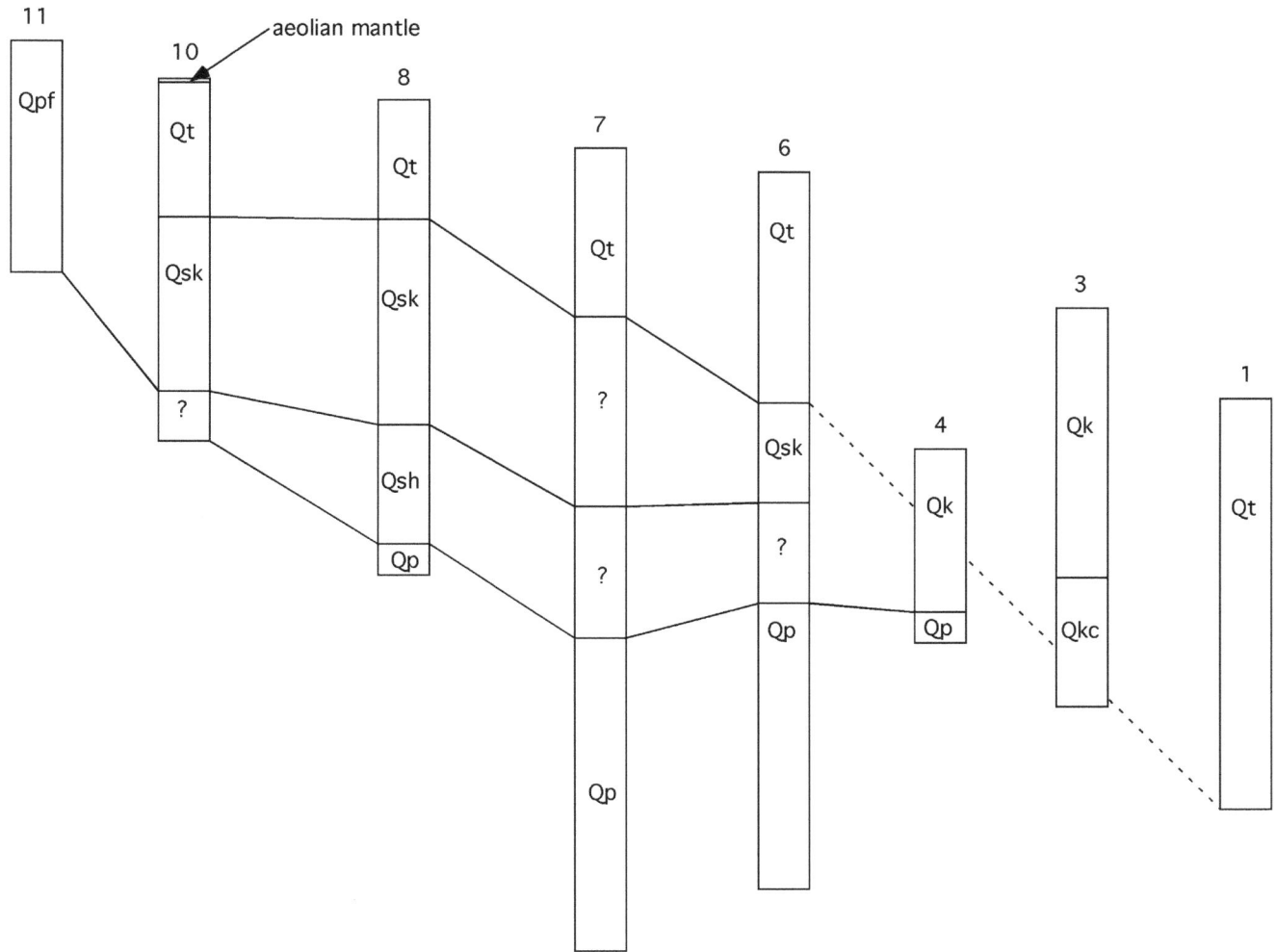

Figure 8. Correlation of stratigraphic units between selected profiles. Dotted lines show less certain correlations. Question marks (?) indicate alluvial units not positively identified, though their origin may be inferred by the stratigraphic correlations shown. The relative elevations of each profile are not to scale (see text).

narrow the age range of sites already discovered. The geological data can be used to place sites into the "ages" used in East Africa that are broadly characterized by technological aspects of the toolkits represented by certain time periods. From youngest to oldest, these are the Iron Age, Later Stone Age, Middle Stone Age, and Early Stone Age.

Iron-smelting technology has been practiced around Lake Victoria in Kenya for at least 2,000 years and perhaps up to 1,000 years earlier (Schmidt 1975; van Grunderbeek et al. 1988). It is unclear when such technology may have diffused to, or emerged, in Central Kenya. Iron Age technology compliments and then replaces a microlithic Later Stone Age.

The Peleta, Kipsing, and Shordika surfaces have been exposed since the early Holocene, so Iron Age sites might be found on these surfaces, but never buried within them. Since it is likely that Tol alluvium has buried older surfaces, archaeologists must be aware that Iron Age sites might be found in a buried context between Tol alluvium and any underlying unit.

Iron Age sites, then, might be found on the surface of the Tol alluvium, buried in the Tol alluvium, on the surface of the Seaku paleosol, or buried within the Seaku alluvium. Just as today, many older alluvia have a surface expression in the project area, and Iron Age peoples may have been responsible for archaeological sites on those surfaces.

The East African Neolithic period, characterized by pastoralism in the study area, is a subject of great interest. One of the major goals of ongoing archaeological research is to determine the timing of the introduction of pastoralism into the project area. The earliest commonly accepted dates for the introduction of domesticated animal herding in Kenya come from the northern regions of Kenya. Sites in the Ileret area on the northeastern shore of Lake Turkana show that pastoralism was established there between 4000 and 5000 B.P. (Barthelme 1985; Marshall et al. 1984; PT Robertshaw and DP Collett 1983). The pastoral Neolithic is part of the Later Stone Age, but foraging was also a common economic adaptation in the Mukogodo Hills until very recently. Therefore, not all stone age sites discovered in Tol and Seaku alluvia are necessarily associated with pastoralists.

Table 2: Predictive Site Model

Cultural Period	Quaternary stratigraphic unit						
	Qt	Qsk	Qsh	Qk	Qkc	Qp	Qpf
Historic	○	○	○	○	○	○	○
Iron Age	●	●	○	○	○	○	○
Neolithic	-	●	○	○	○	○	○
Later Stone Age	-	-	●	○	○	○	○
Middle Stone Age	-	-	-	●	●	●	●
Early Stone Age	-	-	-	✓	✓	✓	✓

○   Sites on surface possible
●   Buried sites possible (also may be on surface)
-    Not possible unless in secondary context
✓   Undetermined

In the current study area, the Neolithic is represented by the deposition of the Seaku alluvium and the expansion of the grassland environment indicated by the Seaku Paleosol. Later Stone Age sites might be found on the surface of the Seaku paleosol or buried within the Seaku alluvium. Both the Peleta and Shordika surfaces have been exposed since the early Holocene, so Later Stone Age sites might be found on their surfaces, but never buried within them.

The Shordika alluvium was deposited during the late Holocene, yet prior to the introduction of pastoralism in East Africa. Therefore, only Later Stone Age sites should be found buried within its alluvium. Its surface has been continuously exposed throughout the Holocene. Wherever the Shordika surface is exposed today, it might contain Later Stone Age through Iron Age sites, and even Historic period sites. In a few locations the Shordika alluvium is preserved in a buried context. Sites on that buried surface contact would have a more limited chronological expression determined by the age of overlying sediments. All sites on the Shordika surface should be less than about 8,500 years old.

The Peleta and Kipsing alluvia are the oldest lithostratigraphic units in the study area. Their maximum ages are as yet undetermined, but they both show characteristics of having undergone substantial soil formation processes during the hyper-arid conditions of the Pleistocene. Because both have been exposed to the surface for long periods of time, Middle and Later Stone Age, as well as Iron Age and Historic period sites, are likely to be discovered on its surface. However, because of the antiquity of these geologic units, only Middle Stone Age and possibly Early Stone Age sites are likely to be found in a buried context. Since there are several locations in the study area where the Peleta appears as a relict surface (i.e. a surface that has been re-exposed through erosion), it is possible to find young archaeological deposits lagged onto its surface.

All time periods are represented in the study area. That is, erosion has not completely obliterated any one period, with the possible exception of early Pleistocene deposits that contain evidence of pre-*Homo sapiens* development and Early Stone Age technology. Additional study may reveal whether or not sediments representing those time periods have been completely flushed out of the region by erosion, are preserved buried somewhere beneath the Peleta alluvium, or if in fact the Peleta surface itself was subject to occupation during the early Pleistocene.

Another of the major goals of the current study is to determine to what extent archaeological site distribution and preservation have been affected by ongoing natural site formation processes. One of the first observations made by archaeologists in the study area is that Middle Stone Age sites are differentially distributed on the Quaternary surfaces. In particular, Middle Stone Age sites are encountered only where the Peleta alluvium is exposed. Later Stone Age sites, however, are poorly represented on the Peleta surface, even though it was available for use and occupation during the Later Stone Age.

The reason that Middle Stone Age sites are not encountered on Shordika and younger surfaces is that these sediments are too young to have been used by Middle Stone Age peoples. Furthermore, some Peleta sediments have been eroded away and either flushed out of the valley completely or incorporated into younger sediment deposits downstream. Explaining the uneven distribution of Later Stone Age sites is more difficult, but landscape change can be ruled out as a contributing factor. The significance of this is discussed in the following chapter.

CHAPTER 4
SITE SURVEY

In 1996 and 1999, in conjunction with the National Museums of Kenya, a large-scale pedestrian survey of the Tol River and lower Seaku River floodplains was conducted. Seventy-one new archaeological sites (or isolated finds) were recorded over an approximate 24 km^2 area (Figure 9) using the techniques described below. The primary objective of the survey was to learn as much as possible about the archaeology of the area surrounding Kakwa Lelash and Shurmai rockshelters. Both Shurmai and Kakwa Lelash are believed to have been semi-permanent habitation sites used intermittently throughout the Late Quaternary. A complete record of human activity also includes nonhabitation sites; i.e., sites possibly used by inhabitants of those caves in their day-to-day activities. Furthermore, modern inhabitants of the region temporarily settle in open-air sites and use the rockshelters as special-use sites. Thus, the pattern of site use is apparently reversed in the modern setting. We do not know when the contemporary settlement pattern began, nor do we know if it was a slow or rapid transition. Elucidating the details and timing of this transition are important goals of our research program. Site survey is an important step in a research program designed to address regional archaeological problems such as these.

METHODS

Each site was fully described on a detailed site form and exact Universal Trans Mercator (UTM) coordinates were obtained using a portable Global Positioning System (GPS). As a backup, each site was plotted onto a regional field map (1:50,000) using a surveying compass. All surface artifacts and samples collected in the field were taken to the National Museums of Kenya (NMK) for analysis and curation. As per NMK procedure, each site was given a unique identification code to replace the temporary field identification number. It is NMK policy to assign site numbers sparingly. Some sites that were initially questionable were assigned "spot find" numbers. These are indicated in the text with the prefix of "0/" with their site number. Though these sites are usually small, they are still significant. As such, they are included in the overall analysis of land-use patterns.

The number of sites located in reconnaissance is directly proportional to the intensity with which that reconnaissance is conducted (Plog et al. 1978:389-394). The question facing the archaeologist is what level of intensity is necessary to accomplish the research goals? In the Kipsing area there are a number of factors that must be considered, including visibility and accessibility of sites. The landscape within the project area is highly eroded due to long-term utilization by nomadic pastoralists. Vegetation is limited and dispersed, leaving ground visibility high, even during the wet season. With such high visibility, 30 m transect intervals (15 m per side) are sufficiently narrow to allow each surveyor to view the full transect width. Even so, at this intensity some unobtrusive sites might be missed. This would include extremely small or sparse sites. However, Schiffer and Wells (1982:347) note that in order to discover sites of low obtrusiveness, it would be necessary to survey at impractical intensities; therefore, the target survey intensity is appropriate.

In order to provide a package of data that fit well with previously collected information, connecting Shurmai and Kakwa Lelash rockshelters with survey coverage was an important research objective. A systematic sampling strategy was employed in order to increase the maximum geographic extent of the survey. Approximately 30 percent of survey area was covered by systematically surveying from one end of the survey area and leaving about 300 m between return transects.

The Quaternary formation processes of the sedimentary record in the study area are just now being understood. As discussed in the previous section, geoarchaeological investigations have demonstrated that significant deposition and erosion have taken place during the late Quaternary and Holocene along the floodplain of the Tol and Seaku Rivers. It is likely that some archaeological sites have been buried and preserved in the sediments. Indeed, previous survey had shown that some isolated finds could be found in deeply buried sediments exposed in stream cuts. Mechanical subsurface testing was logistically impossible, and manual subsurface testing would have been both inefficient and ineffective given the 10-15 m depth of floodplain deposits. Instead a survey design was implemented that maximized surface coverage and natural exposure examination. Over 16 linear km of natural exposures in the Tol and Seaku drainage networks were inspected for subsurface cultural deposits. We also inquired with local residents about their knowledge of potential archaeological sites, especially rockshelters that are more readily recognizable.

Because so little is known about the archaeology of this region, because data are sparse, and because these sites may never be visited again, a decision was made to surface collect each site. Under these conditions, "the potential for data enrichment through artifact collection generally outweighs the need to conserve contextual data" (Bower 1986:25). However, this decision was not made lightly. Sampling was not generally an issue because most sites were so small that complete collection was the norm. For larger sites, an arbitrary representative sample was collected. The goal of collection was to adequately characterize each site in terms of its composition.

The methods proposed here are similar to those reported for the Lemek-Mara area of Kenya (Robertshaw et al. 1990). Survey in the Lemek-Mara area benefited from a preexisting culture-historical framework within which to more succinctly classify sites. The development of such a framework is one of the goals of this publication.

PROBLEMS ENCOUNTERED

Generally speaking, survey conditions in July and August in Kenya are quite good. The weather is hot and dry, and

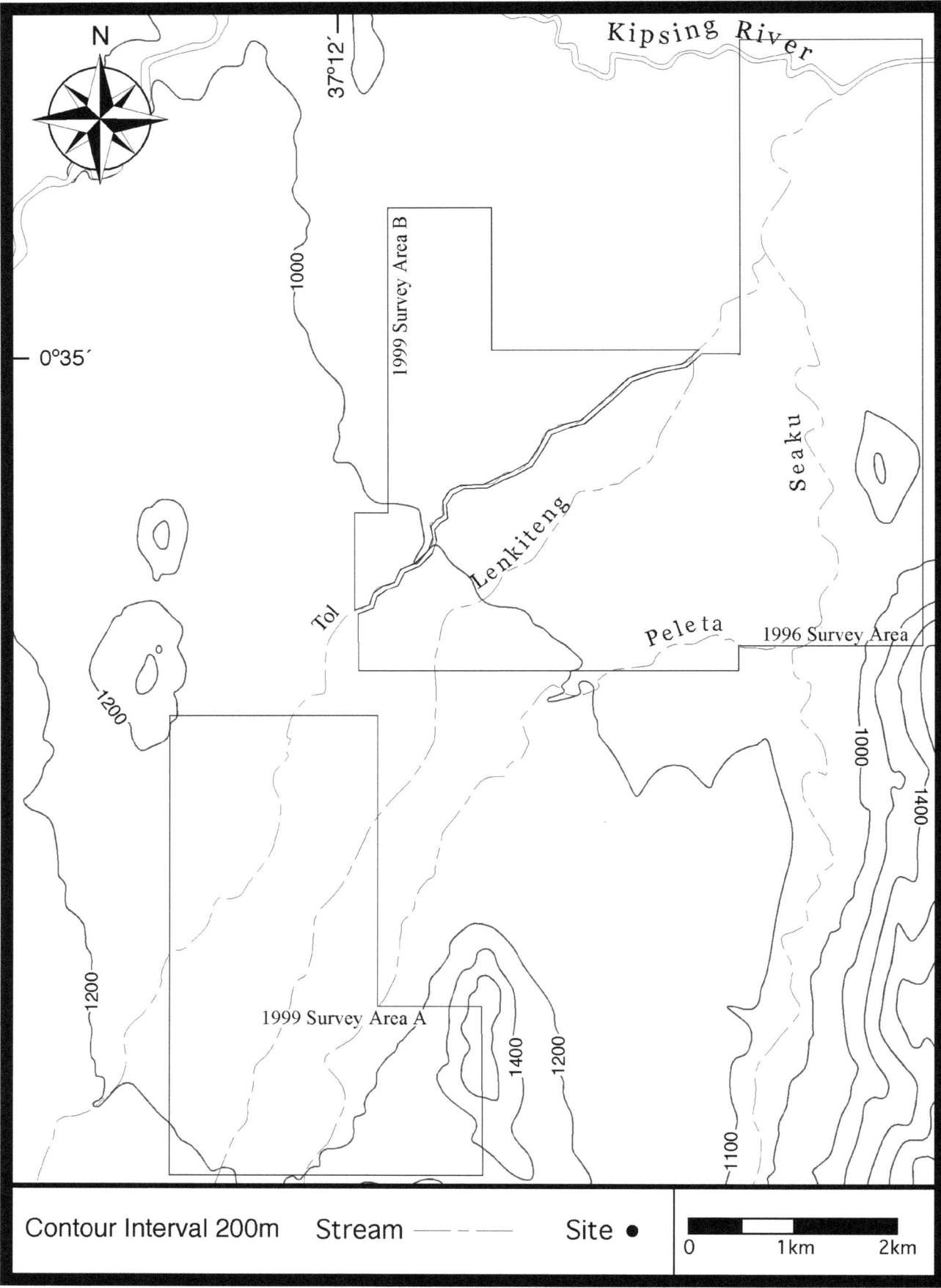

Figure 9. Map of the study area showing boundaries of Survey Areas A and B.

as long as the surveyors get plenty of water and enough food, all is well. Maintaining supplies is always an issue in the remote parts of Kenya.

Just as we arrived in the project area in 1996, ethnic violence broke out and caused us to temporarily abandon our survey efforts. We returned in 1999 when the violence had subsided. Inhospitable terrain and thorny acacia scrub often forced transects to deviate and occasionally to terminate altogether. Overall, however, the 1999 season was successful and we achieved all of our research objectives.

SITE CHRONOLOGY

The primary dating techniques used to arrange the sites in our corpus in relative time was lithic and ceramic seriation, and geochronology. Surface collections from the sites located in our surveys produced 589 ceramic sherds and 2,165 lithic objects. The sherds were analyzed by Dr. Simiyu Wandibba of the Institute of African Cultures of the National Museums of Kenya. The lithic objects were analyzed in the United States at the close of the 1999 field season by Dr. G-Young Gang of Yeungnam University, Korea, as part of her post-doctoral study at Texas A&M University. Dr. Gang conducted the earlier analysis of the lithic assemblage recovered from the stratified deposits at Shurmai and Kakwa Lelash rockshelters, and she has provided us with an accurate picture of the subtle changes in lithic manufacturing techniques and raw material use that occurred in our study area over time (Gang 2001). Similarly, Dr. Wandibba's chronology of ceramic technology in Kenya provided us with a means of cross-dating the later period site occupations. A site's position on the landscape is less precise an indicator, but does allow us to rule out an ancient age for sites on young landforms.

We did not expect open-air sites to have the same distribution of characteristic materials as the rockshelters; indeed, we expected that they would be significantly different. For this reason we identified some key characteristics to help guide placement of the survey sites into a chronological framework.

Early Stone Age (ESA)

No sites from the ESA were found on this survey. Some sites in the survey were initially suspected to be ESA, but upon reconsideration have been placed in the MSA. Early Stone Age stone tools are the oldest stone tools in the world and are found within several hundred kilometers of the study area at such sites as Olorgesailie, Kariandusi, Kapthurin, and Kilombe, to name a few. Because we do not know for certain how ancient the sedimentary geology is in our area it remains conceivable that early stone tools could be found here.

While remarkable Acheulian hand axe would have been very difficult to overlook, an ESA site composed entirely of trimmed flakes might not be so obvious. We would expect late ESA sites to contain larger Levallois-type flakes and cores, to be accompanied by other large generalized flakes and flake tools, and possibly hand-axes and/or choppers assignable to the Acheulian tradition.

Middle Stone Age (MSA)

The age and definition of the MSA is a matter of ongoing debate (see pp. 7-8). The term is usually applied to assemblages of lithic material that exhibit the presence of the Levallois or prepared core technique (Phillipson 1993:60). This technology produced numerous specialized blades, backed-blades, uniform triangular flakes, Levallois flakes, and formal core types (Ambrose 2001:1751; Clark 1970:124; McBrearty and Brooks 2000:495). The MSA has been dated in East Africa between 200,000 and 40,000 B.P. (Robertshaw 1995). The transition between the MSA and subsequent Later Stone Age is not as distinctive as once thought, as is clearly reflected in the gradual change in lithic technology at Shurmai and Kakwa Lelash rockshelters (Gang 1997:258-262). Although this makes it difficult to assign dates to lithic assemblages based solely on technological, morphological, and stylistic attributes, however, four principle changes were noted:

1. Procurement strategy shifts from primarily locally abundant and locally available raw materials in the earliest levels, to roughly equal proportions of local and nonlocal materials in the uppermost levels.
2. There is a marked change in the types of flakes used as tools, with larger, longer flakes and attendant increased platform thickness in the earliest levels.
3. There is a significant decrease in artifact size and concomitant core size reduction. Cores are more likely to be exhausted as knappers maximize use of nonlocal resources.
4. Production of blades and backed knives declines, such that there are very few blades and no backed knives in the upper levels.

The deepest stratigraphic level containing artifacts at Shurmai Rockshelter (the deeper of the two excavated sites) yielded an infrared stimulated luminescence minimum age of 45,211 ± 5,356 B.P. An AMS radiocarbon age of 20,000 ± 80 B.P. was obtained from organic material in the immediately overlying sediments. The earliest associated lithic artifacts are thus characterized as MSA by Dickson and Gang (2002). The uppermost layer, containing material that was radiocarbon dated to 1,100 ± 90 B.P. (Kuehn and Dickson 1999:80), is known to have been occupied historically (Cronk 1989b).

Based on research done by Gang (2001), several key indicators of MSA sites should be discernable. First, an MSA site must be dominated by basalt as the principle raw material for lithic manufacture. Lithic debitage in the lowest level of Shurmai consisted of 92.44 percent locally available material with the remainder being nonlocal or exotic (Gang 1997:192). This value dramatically declines in the uppermost level to only 18.42 percent. At Kakwa Lelash, an entirely Later Stone Age site, proportions of local to nonlocal materials are approximately 1:1. Though site formation processes associated with open sites are undoubtedly quite different than those seen at rockshelters, raw material ratios represent the first clue about site age. A preponderance of large basalt cores, and especially basalt core tools, also indicates an MSA component. Additionally, a tendency towards large basalt flakes and a preference for blades strengthens the MSA argument. Figure 10 shows several examples of flakes attributed to the MSA. Long flakes such as these (over 50 mm) were typical at almost every MSA site recorded. Figure 11 shows several large cores attributable to the MSA. All artifacts in these two figures are basalt.

Later Stone Age (LSA, also Late Stone Age)

Later Stone Age sites include all sites dominated by microlithic technology. Some LSA sites containing expedient tools of local materials, or the flaking debris thereof, are also considered. Such sites are differentiated from similar MSA sites by their smaller flake size and few, if any, basalt core tools. Figure 12 shows a sample of LSA flakes seen during survey. Several nonlocal raw materials are shown, including chert variants and obsidian. Most LSA sites with tools are easily identified by the abundance of nonlocal raw materials required for finer control in the manufacturing process, and also by the mixed presence of pottery and lithics. Occasionally, an LSA site was identified by a preponderance of small basalt flakes.

Iron Age (IRA)

The diagnostic characteristic of Iron Age sites is the presence of iron-smelting or smithing debris. Sites with ceramics but little or no lithic manufacturing debris or stone tools are also tentatively considered Iron Age sites. With few exceptions, these latter sites are represented by the numerous isolated finds that contain only ceramics. Occasionally, a thick-walled vessel indicative of early ceramic technology is found and the site is classified LSA. Though ceramic production originated in the LSA, most LSA sites have lithic debris.

DISCUSSION OF SURVEY RESULTS

Seventy-one archaeological sites were recorded during this survey. A list of the sites in our corpus, together with a summary of the analysis of the lithic and ceramic material recovered from each of them is presented in Table 3, and detailed site descriptions appear in Appendix A. Although the possibility seems remote, the specific provenance of each site has been omitted from this discussion in order to protect these archaeological sites from illicit digging or disturbance. Authorized parties can obtain the map coordinates and other provenance information for these site from the site files maintained by the Division of Archaeology, National Museums of Kenya, Nairobi.

The total amount of collected flakes, cores, tools, and other lithic debitage is given in Table 3 to provide the reader an idea of the collection size at the National Museums of Kenya.

Using seriation and geomorphological position, we recognized eight types of sites in our site corpus: rockshelter sites, lithic scatters with Middle Stone Age (MSA) morphological characteristics, lithic scatters with Later Stone Age (LSA) morphological characteristics, scatters of both LSA lithics and ceramics, scatters of ceramics, iron smelting sites, rock cairns, and Colonial period and other sites. Some sites have multiple components, with each component listed if possible. The geomorphic context of each site is also given. Since the Peleta fan (Qpf) and Peleta alluvium (Qp) are difficult to distinguish, they are grouped on Table 3. Likewise, Kipsing alluvium (Qk) and Kipsing calcrete (Qkc) are grouped. Only one site, GnJm 47, was both multicomponent and had multiple contexts. In this case, the Shordika alluvium is eroding to show the Peleta unit. The MSA component appears to be coming from the Peleta alluvium.

Stone Age Sites

Most sites discovered during survey are referable to either the Middle or Later Stone Age. The MSA occupants of the region do not appear to have had a varied land-use strategy (Figure 13). The only place MSA sites are found is on Late Pleistocene surfaces. Differences in site distribution appear to be the result of natural site formation processes rather than intentional land-use strategies. The density, distribution, and composition of MSA sites do not vary significantly where these sites are preserved.

The majority of the LSA sites discovered are located at the southern end of the survey area (Figure 14). Indeed, they are conspicuously absent from the northern survey area around the lower Seaku River. However, the LSA is nicely represented at Kipsing by the multicomponent Kipsing site (GnJm 23). Later Stone Age sites are most abundant in the Mukogodo foothills.

Generally speaking, the Late Pleistocene surfaces where MSA sites are encountered would have been available for use and occupation by LSA peoples as well (Table 2). Thus, the difference in site distribution on these surfaces between the MSA and LSA cannot be fully explained by natural site formation processes, and are probably the result of cultural processes. While LSA sites are predominately found in the interfluvial area between the Tol and Peleta drainages as one approaches the Mukogodo foothills, MSA sites are abundant at both ends of the survey area.

Iron Age Sites

A number of Iron Age sites were discovered on survey (Figure 15). Most were ceramic scatters, probably associated with food producers. Two sites, GnJm 15 and GnJm 55, were related to iron production. Historically and archaeologically, iron manufacture in Kenya is the

Figure 10. Examples of Middle Stone Age flakes from GnJm 18. (a) Ventral and (b) dorsal views.

Figure 11. Examples of Middle Stone Age cores from GnJm 23. (a) dorsal view of core tool, (b) ventral view of same, (c) top, and (d) bottom views of another core.

a

b

Figure 12. Examples of Later Stone Age technology from GnJm 47. (a) ventral and (b) dorsal views.

Table 3. Corpus of Archaeological Sites Recorded During the Mukogodo Hills Site Reconnaissance.

Site #	Site Type	Period(s) of Occupation	Depositional Context	Flakes	Cores	Tools	Flaking Debris	Sherds
GnJm 7	Lithic Scatter	Early MSA	Peleta	16	24	10	59	—
GnJm 8	Stone Cairns	?	Peleta	—	—	—	—	—
GnJm 9	Ceramic Scatter	IRA	Tol	—	—	—	—	<25
GnJm 10	Ceramic Scatter	IRA	Tol	—	—	—	—	<25
GnJm 11	Ceramic Scatter	IRA	Tol	—	—	—	—	<25
GnJm 12	Lithic Scatter	MSA or LSA	Peleta	69	15	3	172	—
GnJm 13	Lithic Scatter	MSA	Peleta	26	10	6	73	—
GnJm 14	Lithic Scatter	MSA	Peleta	15	5	5	21	—
GnJm 15	Iron smelting site	IRA	Seaku	—	—	—	—	—
GnJm 16	Lithic Scatter	MSA	Kipsing	25	7	4	39	—
GnJm 17	Stone Cairn	?	Kipsing	2	1	4	9	—
GnJm 18	Lithic Scatter	MSA/IRA	Kipsing	11	7	—	8	—
GnJm 19	Lithic Scatter	LSA	Kipsing	6	1	—	25	—
GnJm 20	Stone Cairn	?	Kipsing	—	—	—	—	—
GnJm 21	Stone Cairn	?	Peleta	—	—	—	—	—
GnJm 22	Lithic Scatter	MSA	Peleta	13	8	11	27	—
GnJm 23	Lithic Scatter	MSA/LSA/IRA/H	Kipsing	15	12	9	17	—
GnJm 24	Stone Cairns	?	Kipsing	—	—	—	—	—
GnJm 25	Stone Cairns, Lithic Scatter	MSA/?	Peleta	3	1	1	7	—
GnJm 26	Stone Cairn	?	Peleta	—	—	—	—	—
GnJm 27	Ceramic Scatter	LSA	Tol	—	—	—	—	163
GnJm 28	Ceramic Scatter	IRA	Tol	—	—	—	—	3
GnJm 29	Multiple component prehistoric use	MSA/LSA	Peleta	1	—	—	11	—
GnJm 30	Ceramic Scatter	LSA	Peleta	—	—	—	—	4
GnJm 31	Lithic Scatter	MSA	Peleta	29	6	5	45	—
GnJm 32	Lithic Scatter	MSA/LSA	Peleta	18	1	2	33	—
GnJm 33	Lithic Scatter	?	channel	—	—	—	2	—
GnJm 34	Lithic Scatter	MSA	Kipsing	—	5	4	13	—
GnJm 35	Ceramic Scatter	Historic	Peleta	—	—	—	—	98
GnJm 36	Lithic Scatter	MSA	Peleta	17	1	1	54	—
GnJm 37	Lithic Scatter	MSA	Peleta	2	5	—	19	—
GnJm 38	Buried Lithic Scatter	LSA	Seaku	9	—	—	25	—
GnJm 39	Ceramic Scatter	IRA	Peleta	—	—	—	1	19
GnJm 40	Lithic Scatter	MSA	channel	—	—	1	—	—
GnJm 41	Lithic and Ceramic Scatter	LSA	Shordika	—	—	—	X	2
GnJm 42	Lithic Scatter	Late MSA	Peleta	17	3	1	39	—
GnJm 43	Lithic Scatter	LSA	Shordika	6	—	1	57	—
GnJm 44	Lithic and Ceramic Scatter	LSA	Shordika	5	—	—	13	10
GnJm 46	Lithic and Ceramic Scatter	LSA	Shordika	—	—	1	—	34
GnJm 47	Lithic and Ceramic Scatter	LSA/H/MSA	Shordika/Peleta	24	—	3	84	10
GnJm 48	Lithic Scatter	LSA/H	Shordika	—	—	—	4	—
GnJm 49	Lithic and Ceramic Scatter	LSA	Shordika	—	—	1	—	16
GnJm 50	Lithic Scatter	?	Peleta	2	—	1	7	—
GnJm 51	Lithic Scatter	Late MSA	Peleta	14	2	—	22	—
GnJm 52	Ceramic Scatter	IRA	Peleta	—	—	—	—	129
GnJm 53	Mau-Mau camp	H	Peleta	—	—	—	—	—
GnJm 54	Lithic Scatter	MSA	Peleta	9	3	1	2	—
GnJm 55	Iron Smelting Site	IRA	Peleta	—	—	—	—	7
GnJm 56	Police Post	H	Shordika	—	—	—	—	—
GnJm 57	Lithic Scatter	MSA	Peleta	2	—	1	7	—
GnJm 58	Buried Lithic Scatter	MSA	Peleta	2	2	2	12	—
GnJm 59	Lithic Scatter	LSA	Kipsing	5	—	—	10	—
GnJm 60	Lithic Scatter	?	Kipsing	1	1	—	—	—
GnJm 61	Lithic Scatter	?	Kipsing	1	—	—	2	—
GnJm 62	Lithic Scatter	MSA	Kipsing	9	3	—	6	—
GnJm 63	Ceramic Scatter	Historic	Kipsing	—	—	—	—	29
GnJm 64	Lithic Scatter	MSA	Kipsing	7	—	—	31	—
GnJm 65	Lithic Scatter	MSA	Kipsing	3	—	1	4	—
GnJm 0/10	Isolated lithic flake	?	Peleta	1	—	—	—	—
GnJm 0/11	Isolated lithic flake	?	channel	1	—	—	—	—
GnJm 0/12	Isolated lithic scatter	?	channel	3	1	—	2	—
GnJm 0/13	Ceramic Scatter	IRA	Peleta	—	—	—	—	5
GnJm 0/14	Ceramic Scatter	IRA	Shordika	—	—	—	—	32
GnJm 0/15	Isolated Lithic	?	Shordika	—	—	—	1	—
GnJm 0/16	Isolated lithics	?	Peleta	—	1	1	—	—
GnJm 0/17	Isolated lithics	?	channel	1	1	1	4	—
GnJm 0/18	Lithic and Ceramic Scatter	LSA	Shordika	—	—	—	1	3
GnJm 0/19	Ceramic Scatter	IRA	Shordika	—	—	—	—	8
GnJm 0/20	Ceramic Scatter	IRA	Kipsing	—	—	—	—	1
GnJm 0/21	Ceramic Scatter	IRA	Shordika	—	—	—	—	16
GnJm 0/22	Isolated lithic flake	?	channel	1	—	—	—	—

Chapter 4: Site Survey

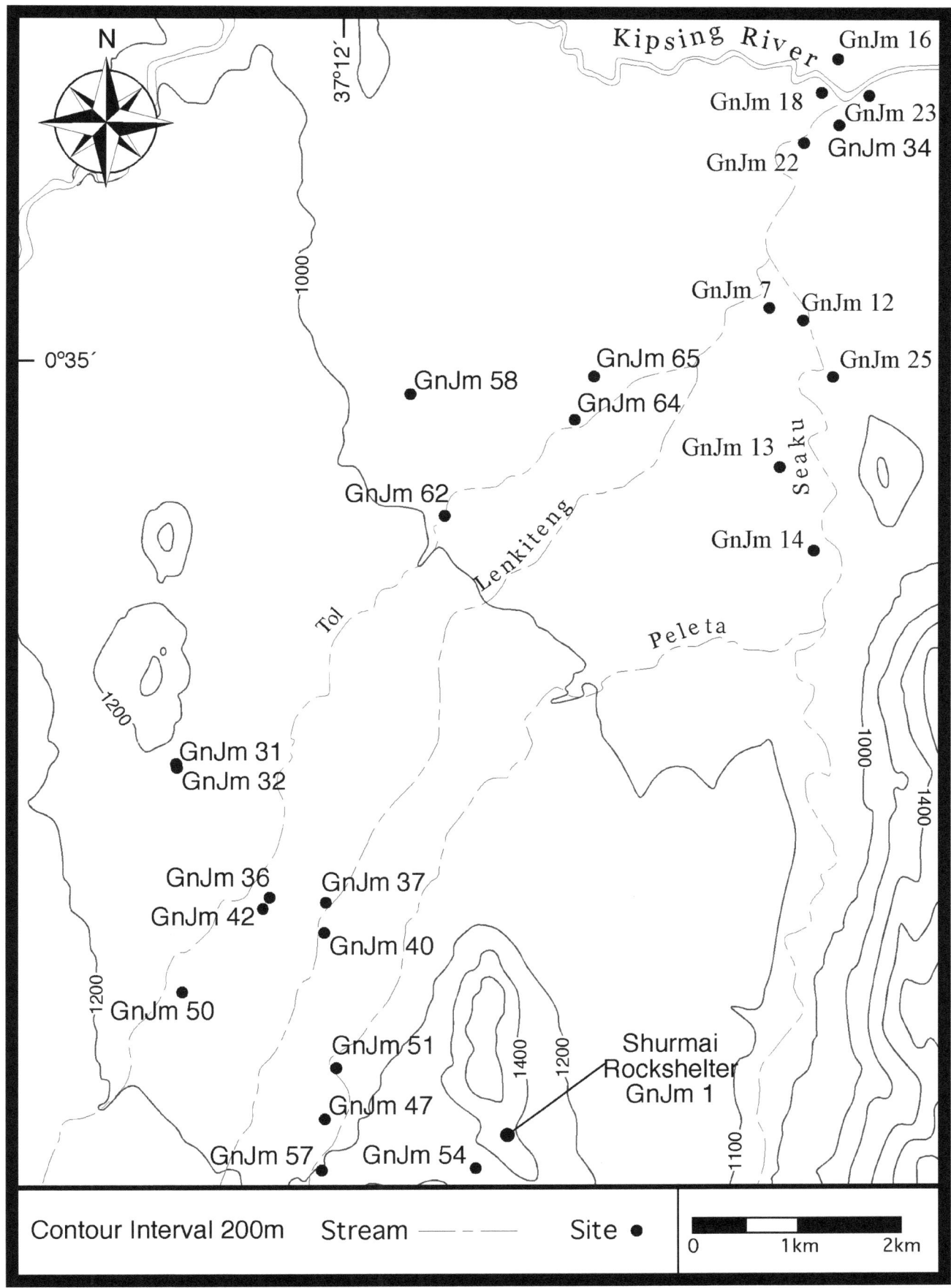

Figure 13. Distribution of Middle Stone Age sites discovered on the survey.

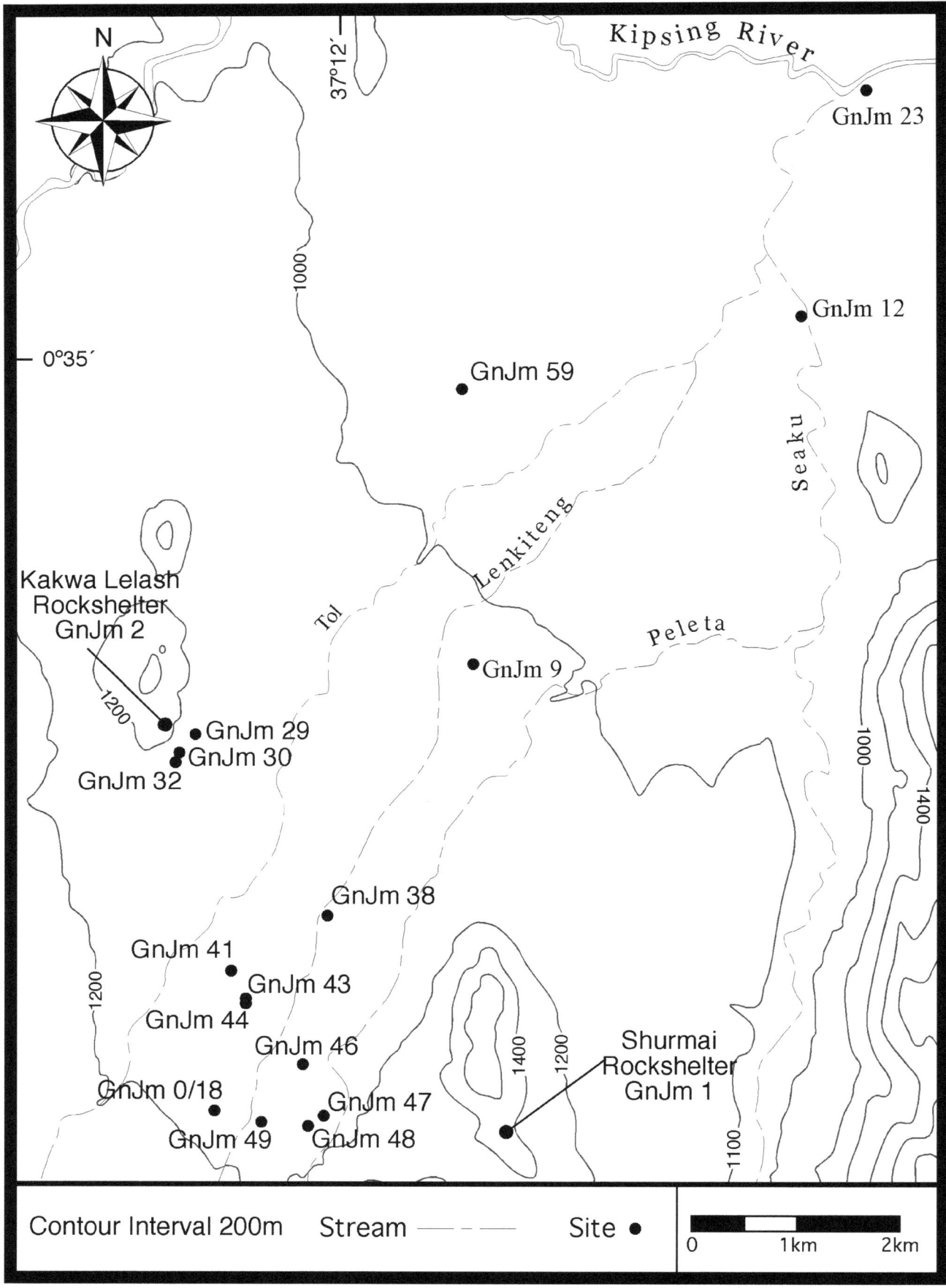

Figure 14. Distribution of Late Stone Age sites discovered on survey.

Chapter 4: Site Survey

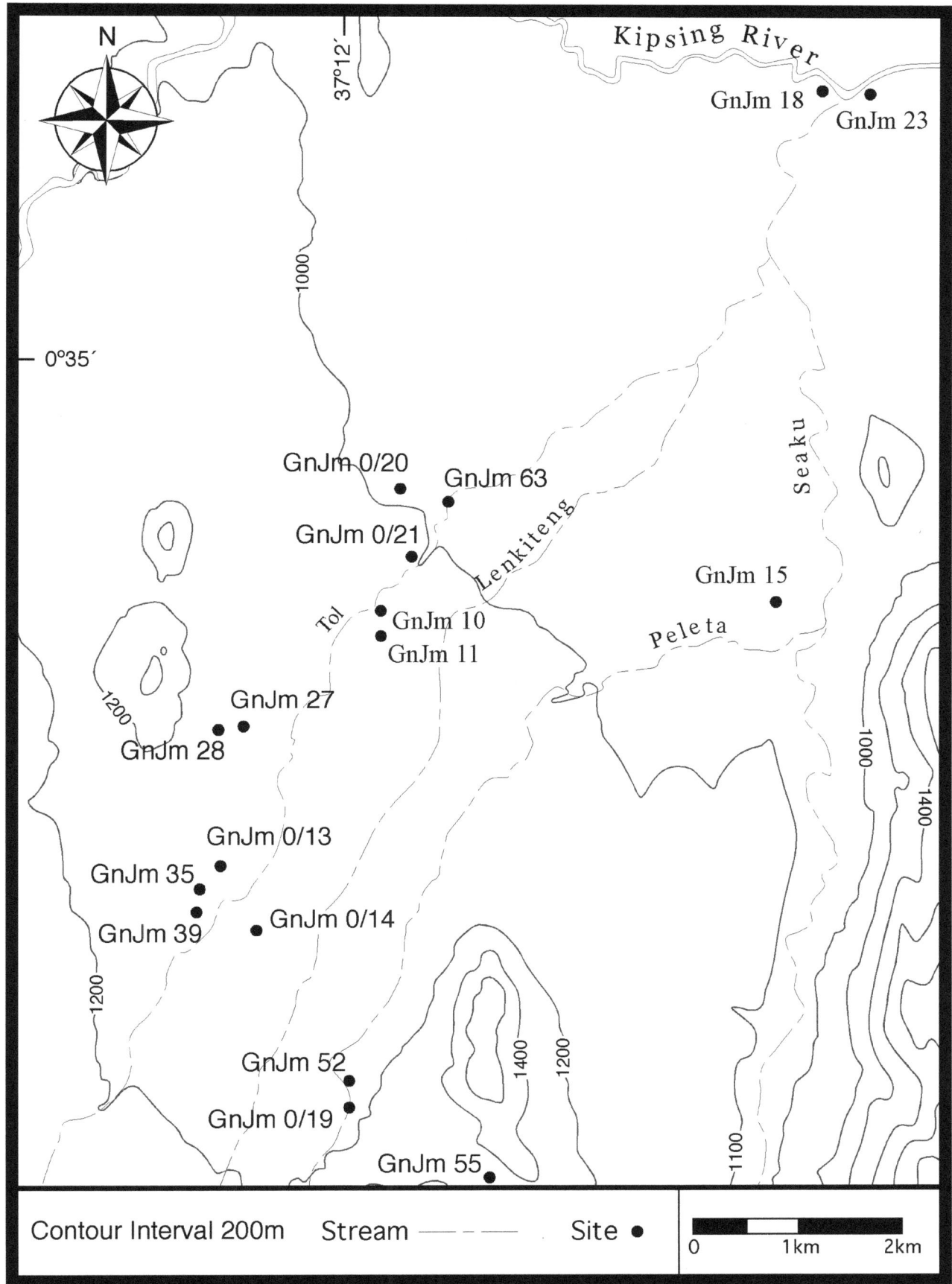

Figure 15. Distribution of Iron Age sites within the study area.

exclusive domain of food producers. The earliest iron use in Kenya is attributed to the Urewe complex of southwestern Kenya, about 2,500 years ago (Phillipson 1993:188). Potentially, iron could have found its way into central Kenya any time thereafter, though it is still unclear when iron became the predominant or exclusive material type (over stone). Both GnJm 15 and 55 contained iron slag, a byproduct of iron production, but only GnJm 55 had the remains of an iron kiln. A thorough excavation of the site remains to be carried out, but it does not appear that GnJm55 was intensively used, and thus was probably not a major iron-producing center. Rather, this site probably represents a supplemental, or even experimental, source.

Iron Age sites have a broad distribution in the project area, similar to the distribution of modern-day pastoralists. It should be noted that stone technology may have been employed very late by foragers who continued to utilize the rockshelters and upland areas. While they undoubtedly would have valued iron tools, much like they had previously prized obsidian and chert tools, they might have continued to make expedient tools from locally available materials. All cultural levels at Shurmai and Kakwa Lelash Rockshelters have a lithic artifact component.

Stone Cairns

Seven sites with stone cairns were observed during survey (Figure 16). Most were found without any associated cultural material. One site, GnJm 25, did have a limited number of basalt and chert flakes and tools. Stone cairns were never discovered far from the seasonal streams that cut across the landscape. Furthermore, all the cairns were found either along the Kipsing or the lower Seaku rivers. At the junction of the Tol and Seaku, a great many cairns of varying sizes were found. A total of 12 cairns was found in sites GnJm 20 (west side of Tol), GnJm 8 (interfluvial area between Tol and Seaku), and GnJm 21 (east side of Seaku). The largest of these sites is GnJm 8, which contained seven of the 12 cairns.

The reason for the clustering of cairns at the junction of the two rivers remains unknown, but it is plausible that the preponderance of cairns along the lower Seaku and Kipsing rivers is due in part to the availability of cobbles in those riverbeds. Cobbles are not typically found in the Tol channel, and pedestrian survey of about 16 km of riverfront land on either side of the Tol and its tributaries discerned no additional cairns.

Local informants do not have any knowledge of how the large archaeological stone cairns that we found on survey were created. Sutton (1973:47) cites a similar state of local knowledge about stone cairns in the western highlands of Kenya. In one of the more extended treatments of stone cairns, he argues that this lack of local knowledge indicates that stone cairns are at least several hundred years old. A similar argument can be made in the Kipsing area. Sutton suggests, in fact, that "the vast majority [of East African cairns] are burial monuments..." (Sutton 1973:37). Citing a few excavated examples that have produced either domesticated animal remains *(*e.g. Posnansky *1968:185)* or iron artifacts (e.g. Fosbrooke 1957:335), Sutton asserts that most cairns were probably built at the terminal end of the Late Stone Age by food producing peoples (i.e. pastoralists), or during the Iron Age. In other words, excavated evidence has not changed L. S. B. Leakey's (1931) contention that stone cairns were a Neolithic African development.

Some local pastoralists near Kipsing have recently begun to bury their dead in graves and construct stone cairns to mark the spot. Direct observation of several burial cairns built between 1997 and 2000 demonstrated that the size of the cairn was related to the age and importance of the individual buried there. One large stone cairn, about 6 m by 5m, was built in 1997 to mark the grave of an important local Turkana elder. Nearby, five small stone burial cairns, each about 1.5 m by 1 m, were built, each marking the grave of a child buried there during the past three years. All of these graves were for migrant Turkana residents of Kipsing, each of whom received a Catholic burial presided over by the local catechist. Samburu informants, however, report that most "burials" in the area are, in fact, open disposals. In this form of disposal, the body is laid in the open without any grave marker and is rapidly consumed by scavenging animals. This fits the description of the Samburu means for disposal of mortuary remains described by Spencer (1973:107).

Pastoralist neighbors of the Samburu routinely practice burial under stone cairns, so the practice may have been more common in the past. For example, to the north, the standard burial practice of the Rendille is beneath a stone cairn (Spencer 1973:59). For the Maasai to the south, mortuary disposal is similar to that of the Samburu, but they acknowledge that, "In the past, important elders, if they so wished, could be buried under a cairn of stones or in the centre of their village with a sprig of wild olive planted above them" (Spencer 1973:240).

Contemporary cairns are also sometimes built to mark roads or as collections for use as building materials. These small cairns are usually less than 1 m in diameter and are characteristically cone-shaped. Furthermore, the stones comprising these cairns are often smaller. Some are no more than piles of gravel. Such cairns were not recorded on the survey. Regardless of method of manufacture, contemporary cairns fit the same distribution pattern as archaeological cairns. That is, they are in close proximity to either the Kipsing or lower Seaku rivers.

The archaeological stone cairns of the Mukogodo Hills are similar in size to cairns found elsewhere in Kenya. These cairns range from 4 to 7 m in maximum length. Cairns from the western highlands of Kenya range from 5 to 18 m (Sutton 1973:41), while cairns from plains of southwestern Kenya range from about 3 to 7 m (Robertshaw et al. 1990:44). Cairns of the Mukogodo Hills are invariably placed on level or nearly level ground, which contrasts somewhat with cairns from the western highlands where they are often placed on slopes.

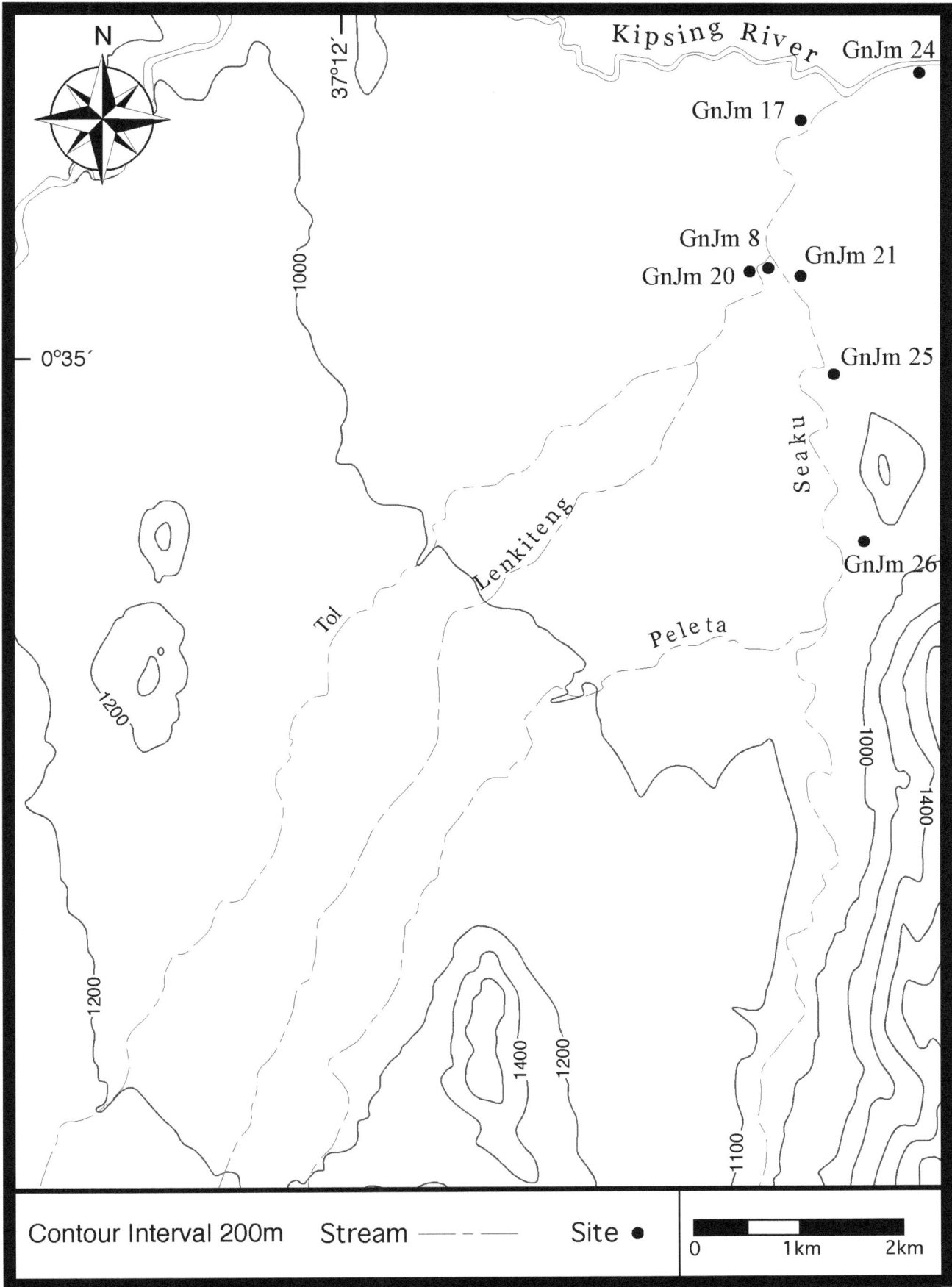

Figure 16. Distribution of stone cairn sites discovered on survey.

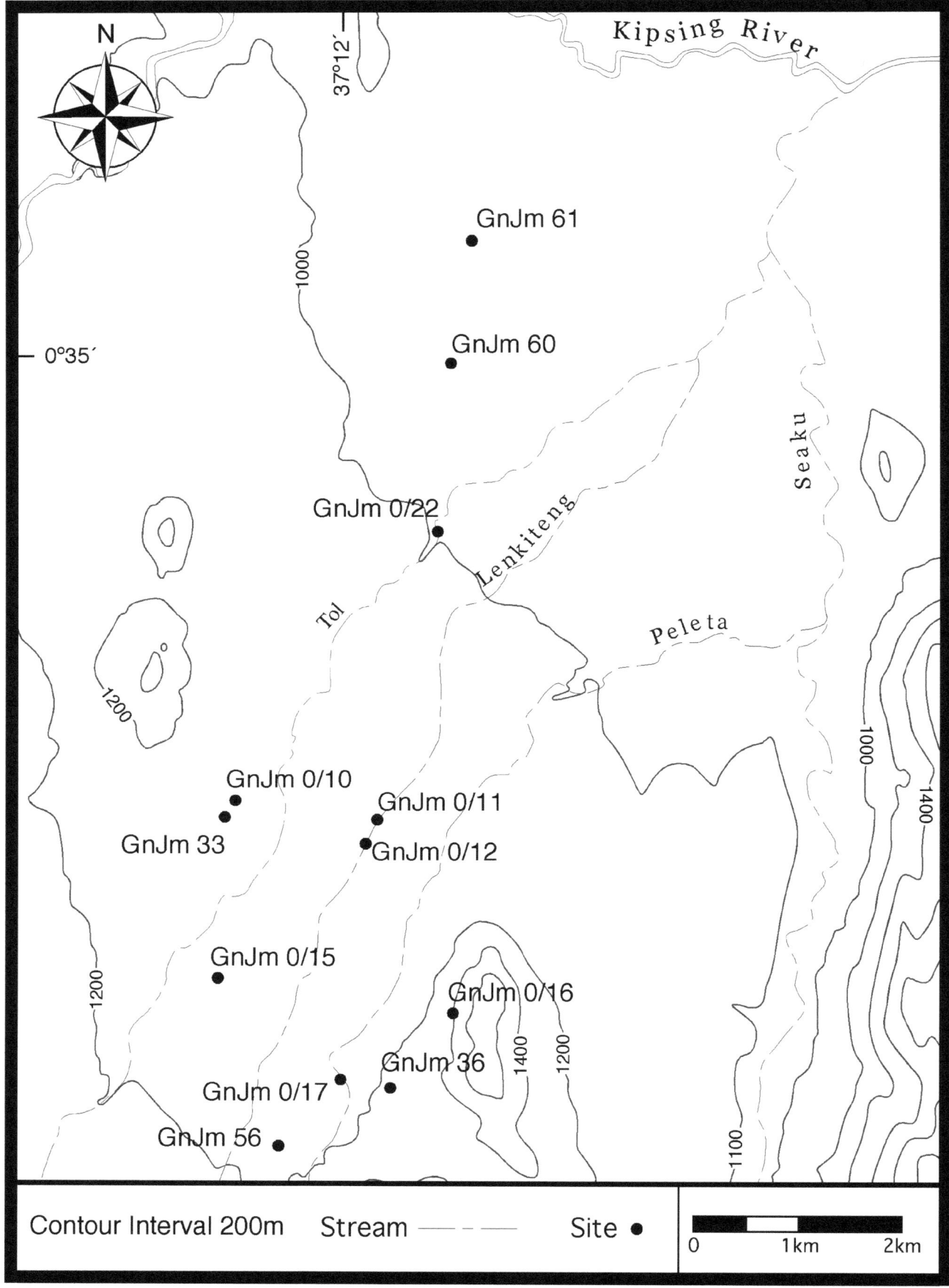

Figure 17. Distribution of other sites discovered on survey.

It is likely that stone cairns in the Mukogodo Hills are in some way related to LSA or IRA peoples' mortuary disposal practices; however, without excavating a few examples, we cannot know for certain. In any case, there are not enough cairns to account for even a very small fraction of the Neolithic occupants of the region. This makes it likely that cairns, if indeed they are mortuary monuments, may in fact represent social stratification.

Other Sites

Numerous other sites were discovered on survey, including isolated finds and historic sites (Figure 17). Isolated finds are sites that are too small or sparse to provide diagnostic information. Historic sites are identified in the study area that played an important role in the historical development of Kipsing, and are often those related to the British occupation of Kenya or Kenyan independence. These sites are less than 50 years old.

CHAPTER 5
SUMMARY AND CONCLUSIONS

REVIEW OF GOALS AND HYPOTHESES

During the course of these investigations, three main objectives were attained. First, we established an alluvial chronology of the Tol River to assist in temporally placing archaeological sites that were discovered in buried contexts. Second, we determined to what extent archaeological site distribution and preservation has been affected by ongoing natural site formation processes. And third, we determined the site distribution pattern of prehistoric sites in the study area. Several hypotheses were tested using independent data collected from pedestrian survey and geoarchaeological investigations in the study area. These hypotheses are now examined in light of data presented in previous sections.

Hypothesis One: Landscape Change (Patterns of Erosion, Deposition, and Stability) Has Not Affected the Distribution of Sites in the Mukogodo Hills-Ewaso Ng'iro Plains

Pedestrian survey efforts demonstrated that the surface distribution of sites of different ages is not even. Sites attributable to the Middle Stone Age were found on the Peleta and Kipsing surfaces, but they were absent from other contexts. Sites attributable to the Later Stone Age were noted on all landforms, but were predominantly found in the interfluvial area between the Tol and Peleta drainages near the Mukogodo foothills. Table 4 shows the actual number of sites discovered in each geomorphic context. Surveyors could not readily distinguish between Peleta fan (Qpf) and Peleta alluvium (Qp), so these units are grouped on Table 4. Likewise, Kipsing alluvium (Qk) and Kipsing calcrete (Qkc) are grouped. Buried sites are shown separately in parentheses. Only sites with a known geomorphic context were tabulated. Sites with more than one context or cultural period were included within each eligible cell. Table 4 provides quantitative data to deduce that representation of cultural periods is not even across the landscape. In fact, since Pleistocene alluvium is largely missing, or buried, in the valley center, this distribution was expected (see Table 2).

Pleistocene alluvium consists of both Kipsing and Peleta units. The distribution of MSA sites is clearly impacted by the erosion of these units, especially the Peleta, alluvium and their subsequent transport downstream. This erosion mainly occurred along the axes of major streams; thus, MSA sites are absent in the interfluvial area where erosion of Peleta alluvium has been significant. The null hypothesis above is therefore falsified. Landscape change *has* affected the distribution of sites in the Mukogodo Hills-Ewaso Ng'iro Plains region.

Indeed, patterns of erosion, stability, and deposition are complex in the study area and must be taken into consideration when interpreting land-use patterns reflected in the archaeological record.

Hypothesis Two: There Is No Significant Difference in Land-Use Patterns Between Middle Stone Age and Later Stone Age Inhabitants of the Mukogodo Hills-Ewaso Ng'iro Plains

Results of the pedestrian survey demonstrated that without a doubt there is a difference in the distribution patterns of MSA and LSA sites. The geoarchaeological survey demonstrated that some of this distribution was not the result of natural formation processes. Table 4 demonstrates that both MSA and LSA sites could occur on exposed Kipsing and Peleta surfaces. However, the majority of discovered LSA sites come from the southern end of the survey and are conspicuously absent from the northern survey area around the lower Seaku River (Figure 14).

With very little exception, wherever MSA sites are found on the surface, LSA sites could also be found. MSA sites are abundant at both ends of the survey area (Figure 13). LSA sites, however, are abundant principally in the Mukogodo foothills at the southern end of the survey area. This contrast in site distribution cannot be explained by natural site formation processes because the Pleistocene surfaces exposed in the northern end of the survey area that contained few LSA sites were also available for use during the Later Stone Age. Thus, it is concluded that there *is* a significant difference in land-use patterns between the Middle and Later Stone Age inhabitants of the Mukogodo Hills-Ewaso Ng'iro Plains, and the null hypothesis is rejected.

There are several possibilities for this disparity. First, it is possible that LSA use of the area was limited, or restricted, to activities that are less visible archaeologically. For example, if LSA peoples were employing a strategy wherein they were logistically tied to the foothill environment, then we would predict an archaeological residue that differs between the foothills and the plains (Binford 1980). Even so, it is hard to imagine literally *no* use of the lower Seaku area by LSA peoples. Rather, the LSA archaeology of that area is fine-grained (*i.e.*, low diversity and sparse).

Table 4. Survey Results Tabulated By Cultural Period And Quaternary Stratigraphic Unit.

Cultural Period	Quaternary stratigraphic unit				
	Qt	Qsk	Qsh	Qk	Qp
Historic	0	0	3	1	1
Iron Age	5	0	4	4	5
Later Stone Age	0	(1)	8	3	4
Middle Stone Age	0	0	0	7	16 (1)
Early Stone Age	0	0	0	0	0

Chapter 5: Summary and Conclusions

The Kipsing site provides an excellent example. GnJm 23 is the modern location of a small settlement of several hundred semi-permanent residents, mostly pastoralists. Bedrock is exposed at the surface and the Kipsing River is entrenched at this nickpoint. Indeed, it is probable that, even though the Kipsing may have meandered somewhat over the course of its history both up and downstream, it has been flowing through this nickpoint throughout the greater part of the Pleistocene. As a result, the Kipsing site has been a place where people have always been able to come to find water. The British identified this locus and built a dam at the nickpoint, but even prior to that time subsurface water would have collected behind the shallow bedrock. Over time this natural subsurface water retention has resulted in salination of the substrate such that groundwater found near Kipsing is slightly salty but potable, a fact not lost on the local pastoralists who see this as a significant benefit to their livestock.

It is no surprise, therefore, that archaeology along the banks at Kipsing includes evidence for more or less continuous human use. Most abundant are the earliest stone tools from the study area, the crude basalt flake and core tools. However, the range of artifact sizes compares to the assemblage from Shurmai Rockshelter, arranged as a palimpsest on the surface at Kipsing. This assemblage includes a small proportion of clearly LSA obsidian and chert flakes, as well as ceramics. These artifacts were not observed during the initial recording of the site precisely because they are unobtrusive. The Kipsing site might best be described as a cluster of overlapping and fused sites of varying ages. The LSA component of the site might have been missed if it had not been for the intense scrutiny given to the recording of the large MSA site component.

Another explanation for LSA site distribution is that LSA people were content to use MSA-style technology in the lower Seaku area, thereby rendering the MSA and LSA assemblages indistinguishable. This, however, seems unlikely as the basalt MSA and LSA assemblages are indeed detectable in stratified deposits at Shurmai. Gang, who conducted both the analysis at Shurmai and of the lithics collected during survey, noted size and stylistic differences that enabled an age estimates. For example, at GnJm 12 there is a preponderance of smaller-sized basalt flakes and a higher proportion of nonlocal lithic raw materials than is typically seen in an MSA assemblage. This site fits neither the MSA or LSA model, but rather, seems to incorporate characteristics of both. Site GnJm 12 demonstrates that it is possible to detect stylistic differences in basalt assemblage composition, and that the bias towards MSA sites in the lower Seaku region is real.

Following Binford (1979), Gang (1997:260) refers to the differences between MSA and LSA raw material use as the difference between *embedded* and *curated* procurement strategies. In turn, I have also followed Binford (1980) and added the concept of *logistically* organized foragers. We are now beginning to see how Stone Age land-use strategies may have changed through the late Pleistocene.

During the MSA, locally available raw materials were principally employed in the construction of a generalized toolkit. This strategy was embedded in basic subsistence schedules; that is, the manufacture of stone tools was incidental to the food procurement activities of MSA people. Though Shurmai Rockshelter is a unique MSA site in the area because it is a rockshelter, the MSA assemblage there is unremarkable in the sense that the interassemblage variability is similar to that at GnJm 7 or GnJm 23. If this means that similar activities were being carried out at both rockshelter and open sites, then it is likely that the MSA peoples were opportunistically exploiting parts of the environment, moving on to new locations when currently available resources were depleted. This implies that the region might have been intermittently occupied. In any event, MSA peoples were likely to have carried on a broad range of activities at open sites.

In contract, LSA sites indicate a different adaptive strategy. The LSA site at Kakwa Lelash Rockshelter is unique in content and scale. While other sites can clearly be defined as LSA based on morphology, none are so large or have as much interassemblage variability as Kakwa Lelash. This indicates that the LSA peoples were employing a central-place foraging system, wherein they carried out the bulk of their archaeologically visible activities at a residential base. The only such campsites that we have identified are at Kakwa Lelash and Shurmai Rockshelters (including GnJm 29 at the base of Kakwa Lelash). The remaining LSA sites can be categorized as *locations* where extractive activities were carried out (Binford 1980). The LSA peoples had moved off of the plains and into the foothills. Additionally, they changed from an embedded resource procurement strategy to a direct procurement strategy, wherein they either obtained raw lithic materials themselves by ranging further afield, or acquired them through trade with neighbors.

Interestingly, the transition from embedded to direct resource procurement strategies appears to have been gradual (Gang 1997:263), but the shift in settlement pattern seems somewhat abrupt. If, in fact, the shift in settlement pattern had been gradual, then we would probably not have been able to see the change expressed so starkly in the surface archaeological record. The emergence of this new land-use strategy also suggests a trend towards strategic planning or "planning depth" (Binford 1989). This concept, which is usually accompanied by curation, signifies (for Binford) the emergence of modern human thought during the Late Pleistocene, a hypothesis tentatively supported by analysis of the stone tool assemblages from Shurmai and Kakwa Lelash rockshelters (Dickson and Gang 2002). The settlement pattern data contained herein provide supporting evidence that a dramatic change occurred.

The change in technology and settlement pattern may, however, simply reflect an intensification of land use by LSA occupants of the Mukogodo Hills due to increased internal population growth, ecological change, or encroachment by other food-producing cultures. The shift a more intensified LSA mode likely preceded the introduction of pastoralism into the area, and it is logical to hypothesize that population growth was a contributing factor. Ameliorating conditions at the end of the Pleistocene may also have contributed to a shift in land-use patterns. Thus, data presented here cannot falsify the contention that the origin of modern human behavior in East Africa occurred earlier during the MSA (c.f. Brooks 1995; McBrearty and Brooks 2000; Yellen et al. 1995).

The earliest food-producing migrants possibly preceded the adoption of iron technology. These populations could be called Pastoral Neolithic peoples (Ambrose 1984). Therefore, the distribution of LSA sites could represent a mixture of two modes of production - foraging and pastoralism. Normally I would expect this to complicate the archaeological record, and indeed it complicates interpretation. Most lithic studies are aimed at interpreting foraging-related behaviors and site formation processes. That having been said, if pastoralists were heavily dependent upon lithic manufacturing technology and their refuse disposal patterns were similar to their foraging counterparts, then we should see more LSA-type material out in the plains where they presumably were camping and grazing their livestock. The fact that this pattern has not been specifically observed suggests that either migrant pastoralists brought iron with them and were never fully dependent on stone tools, or that their refuse disposal patterns were different. Specifically, these Neolithic pastoralists may not have been manufacturing stone tools, but were acquiring them from elsewhere. It may seem unreasonable to expect that Neolithic pastoralists were not only procuring their completed stone tools from elsewhere, but were carefully curating them over an indefinite period of time. But, considering the scale at which pastoralists migrate, they could acquire finer tools from elsewhere without expending much additional effort. Ironically, this brings about the possibility that central-place foragers occupying the Mukogodo Hills rockshelters were obtaining raw materials and/or completed stone tools from Neolithic pastoralists, in much the same way that they would later acquire iron implements.

Hypothesis Three: The Arrival of Pastoralism Did Not Contribute to Erosion and Degradation of the Landscape

The introduction of pastoralism to the study area represents a major ecological change. Pastoralism is often thought to be associated with environmental degradation, especially in terms of surface erosion due to vegetation loss. This is certainly true of pastoralist regions is East Africa today, but may not have been so in the past (McCabe 1989). We hoped that the alluvial chronology and archaeology of the Mukogodo Hills-Ewaso Ng'iro Plains region would shed light on this issue.

Study of the Tol River stratigraphy clearly shows that grasslands expanded 1,000 to 2,000 years ago. During this time, the Ewaso Ng'iro Plains would have been well-suited for pastoralism. Currently, however, no domesticated fauna or other archaeological evidence that unquestionably indicates pastoralism has been found in the study area.

Therefore there is insufficient data to reject the null hypothesis.

If pastoralism arrived as early in the Ewaso Ng'iro Plains as it does in northern and southwestern Kenya, then it would predate evidence for a grassland indicated by the Seaku River paleosol in profile 8 (Figure 7, soil horizons 3A-C). Indeed, the record would show that after the arrival of pastoralism, grasslands flourished for hundreds if not thousands of years.

There is, however, a significant period of deposition that began between 405 and 1,390 years ago. This deposition was the result of significant upstream erosion, perhaps on the slopes of the Mukogodo Hills or from the Laikipia Plateau. If it is associated with pastoralists, is it possible that they arrived that late? There are several other alternatives. First, pastoralism may have arrived earlier, but only expanded into the uplands later. This is not an unreasonable assumption given that the savanna represented by the lowlands would have been much more productive pastoral land. Even historically the uplands were principally inhabited by foragers. The faunal assemblage at Shurmai Rockshelter demonstrates that these same upland foragers did indeed depend on small stock for part of their livelihood (Mutundu 1999:54; also see Cronk 1989a:55). Nonetheless, it is uncertain when pastoralism became the predominant economic pattern in the uplands, or when pastoral use of the uplands reached such an intensity that significant erosion resulted.

Alternatively, upland erosion may be the result of natural processes, such as fire, drought, or other short-term climatic fluctuations. Pastoralists have been known to start grass fires from time to time in order to increase primary productivity (Baker 1975; Evans-Pritchard 1940; Jacobs 1965). If the erosion that caused alluviation of the Tol watershed was the result of pastoral environmental management techniques, then it was a major cultural catastrophe. Abundant dispersed charcoal has been noted along the Seaku/Tol alluvia contact, which is the source of the datable material for that event.

Another possibility is that upland erosion did not accompany pastoralism, but rather the introduction of iron production in the area. While we may never know when iron was actually introduced into the region through trade, the discovery of GnJm 55, an iron smelting site below Shurmai Rockshelter, indicates that some effort was made to produce iron locally. If iron was produced in earnest, then the demand for charcoal may have resulted in upland erosion.

Perhaps the most likely cause of upland erosion is an extended drought coupled with intensive pastoral utilization of the uplands. Significant erosion might occur if pastoralists were occupying the region at moderately high population levels when precipitation-evaporation ratios tumbled. There is some evidence that an arid "Little Ice Age" occurred in East Africa between 600 and 200 years ago (Hamilton 1982). This is indicated by diatomic evidence that correlates with low lake levels at Lake Victoria (Stager and Johnson 2000; Stager *et al.* 1986:86).

There is no direct evidence that prehistoric introduction of pastoralism into the study area had any physical impact on the landscape. Furthermore, there are too many reasonable alternate hypotheses to make any conclusions concerning the cause of alluviation along the Tol River that we see in late Holocene sedimentary deposits. Given that pastoralism is found both north and south of the Mukogodo Hills prior to 500 B.P., the initial introduction of pastoralism seems to be the least likely cause of sedimentation. The null hypothesis is not rejected.

Hypothesis Four: Ecological Change Resulting from Climatic Change Is Not Correlated with Changes in Economic Patterns Observed in the Archaeological Record

Ecological change, insomuch as it is related to climatic change, has not been linear over the past 50,000 years. Rather, there is about a 30,000-year period of arid cooling before dramatic global amelioration began about 17,000 years ago (see Figure 4). The Mukogodo Hills region was occupied during this period, and our survey data suggest that the Ewaso Ng'iro Plains were utilized as well. Based on changes in the moisture budget, we can say that the most dramatic period of climate change in East Africa occurred between about 9,000 and 12,000 years ago. We can now attempt to see if changes in the archaeological record are correlated with ecological change resulting from climate change.

Using the same data as depicted in Figure 4, Figure 18 shows oxygen-isotope ratios and lake-level data for some East African lakes. Significant cultural data is superimposed onto the chart. Occupation of Shurmai Rockshelter begins by about 45,000 B.P. That age estimate comes from unit 2, the oldest cultural-bearing deposit at the rockshelter. Gang (1997) reports that in unit 2, only artifacts referable to the Middle Stone Age were discovered. Over 92 percent of the MSA lithic debitage is made of locally available raw materials. These materials are of poor quality when compared to the obsidians and cherts that were available nonlocally. The fact that some nonlocal materials (less than 8 percent) were present indicates that MSA people were aware of the nonlocal material. Through experimentation, they were aware of its superior fracture and elasticity characteristics. They presumably made a rational decision to not expend additional effort in its procurement. Concerning the lithics, Gang (1997) writes:

> In sum, the analyses of raw material and the techno-morphological, functional, and stylistic attributes of the stone artifacts recovered from the Shurmai (GnJm 1) and Kakwa Lelash (GnJm 2) rockshelters revealed both differences and similarities in the raw material procurement strategies, morphological attributes of artifacts, and stylistic variations between the MSA and LSA occupations at these two archaeological sites. However, this analysis found little difference in technology and functional attributes of the tools at the two sites although, through time, techniques of manufacture become more refined. As Shurmai Rockshelter (GnJm 1) consists of MSA, transitional, and LSA deposits, the site provides us with very useful information about this poorly known period in East Africa prehistory. Finally the site strongly suggests that the development of the LSA from the MSA and the changes in these lithic industries were gradual (Gang 1997:262-263).

This statement strongly suggests that the dramatic amelioration observed at the end of the Pleistocene is not apparent in the lithic assemblages at Shurmai and Kakwa Lelash rockshelters. If true, then this nullifies the argument that determining the history of environmental change is the first step toward understanding the human prehistory of that region. Gang does notice directional changes in the lithic assemblages over time. Notably, these are a shift from reliance on local to nonlocal raw materials, and an overall reduction in artifact dimensions. However, while these changes are directional, change in the climate record is not. Notably, Figure 18 shows that between about 50,000 and 20,000 years ago, the climate was shifting to cooler and drier conditions. During this same period, MSA artifacts corresponded well within the trajectory described by Gang (1997). The lithic assemblage at Shurmai Rockshelter was reduced from 92.4 percent local raw material to 69.3 percent in terms of the composition of the debitage assemblage (Gang 1997:192). The average dimensions of complete flakes during that period were reduced by about 20 to 30 percent, accompanied by about a 7 percent reduction in platform angle (Gang 1997:212).

Twenty thousand years ago East Africa is at the peak of aridity and cool temperatures, though generally speaking the climates 20,000 and 45,000 years ago are more similar to each other than either is to the climates of the Holocene (Figure 18). Between 20,000 years ago and the present the lithic assemblage at Shurmai Rockshelter changed from 69.3 percent local raw material to just 18 percent in terms of the composition of the debitage assemblage (Gang 1997:192), and the average dimensions of complete flakes were reduced by an additional 15 to 25 percent, accompanied by an additional 4 percent reduction in platform angle (Gang 1997:212).

The post-amelioration trend is insignificantly different, leading Gang (1997:263) to conclude that development of the LSA from the MSA was gradual. This does not falsify Hypothesis 3 as established in the goals of this research. It does falsify the hypothesis that long-term ecological change resulting from climate change is correlated with changes in patterns of lithic technology and morphology that are observed in the archaeological record. Specifically, changing climatic conditions did not cause Pleistocene foragers to reduce the overall dimensions of their lithics.

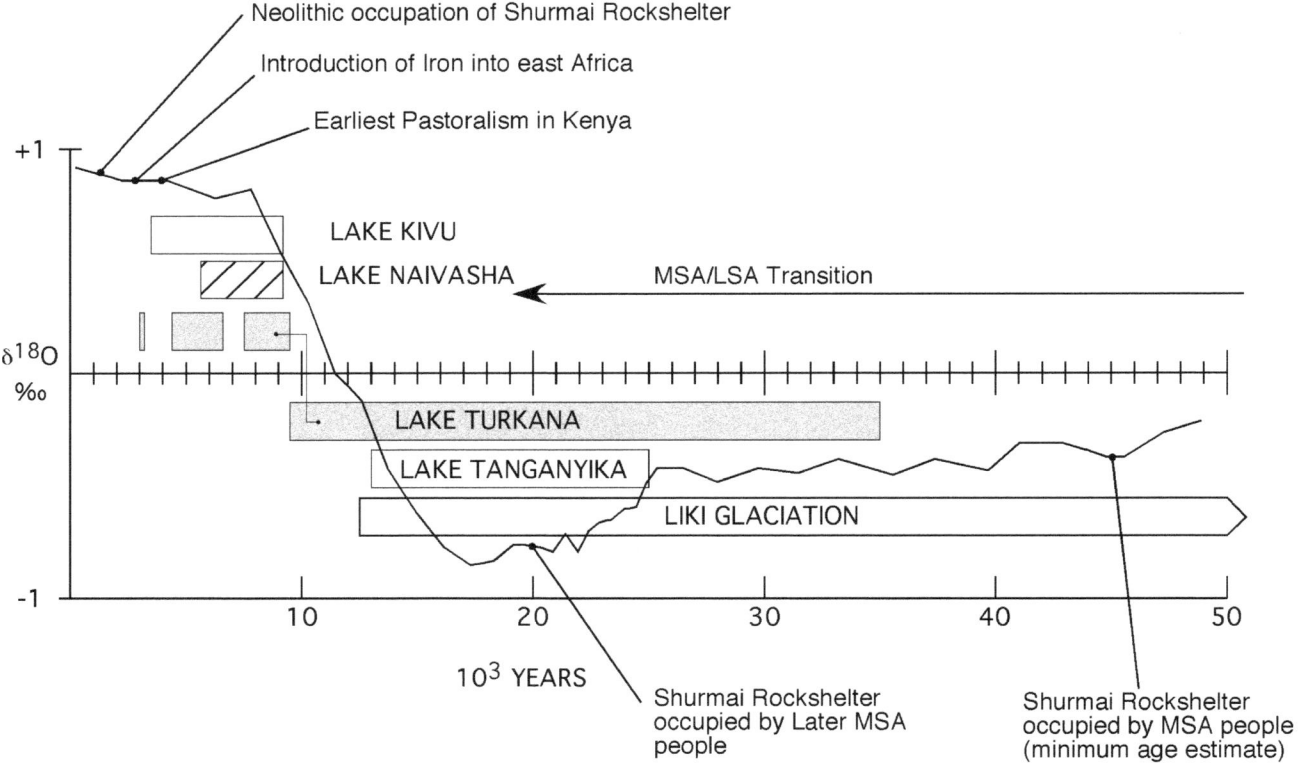

Figure 18. Timeline showing climate change, lake-level fluctuations, and cultural milestones of significance. Oxygen isotope curve based on data in Martinson et al. (1987). Vertical tick marks indicate 1,000 years. Isotope incursions above the line indicate periods of warming and high lake levels. Isotope incursions beneath the line indicate periods of cooling, low lake levels, and glacial advance.

There is still the possibility that changes in economic behavior did, in fact, correlate with climatic change. Other indications of economic change could include either a change in prey choice, a change in procurement tactics, or both.

First let us consider a change in prey choice. The transition from foraging to pastoralism represents a major economic shift and also, in fact, represents a shift in prey choice. However, the period of interest predates the introduction of pastoralism in the area, so we are principally concerned with changes in forager's economic behavior. Unfortunately, these data are not directly accessible based upon the current knowledge set. No time-transgressive faunal material is present at any of the sites. Indeed, only very recent material was preserved in the rockshelters. Provenience of this faunal material is dubious, but it has been reported in Mutundu (1999). In any event, it tells us nothing about long-term changes in prey choice.

It may be possible to model prey choice upon proxy measures, such as changes in technology, but this seems somewhat speculative. It makes more sense to look at other variables in terms of changes in procurement strategies, the analysis of which is linked to prey choice. For example, a shift to pastoralism represents a change in procurement strategy (foraging to herding) and a change in prey choice (wild to domesticated stock). Again, in terms of large-scale climate change, the Pleistocene/ Holocene transition occurs prior to pastoralism, so our focus must temporarily be upon foragers.

Changes in procurement strategies are reflected in the archaeological record. First, as demonstrated at Shurmai and Kakwa Lelash rockshelters, lithic raw material resource procurement strategies change in the Later Stone Age. Though the trajectory of that change was established earlier in the Pleistocene, the scale of change is dramatically accelerated between 20,000 years ago and the present. At Shurmai Rockshelter, there is an over 51.3 percent reduction in the use of local material between the MSA and early LSA, while there is merely a 23.1 percent reduction in the use of local material between 45,000 and 20,000 B.P. (Gang 1997:192). Dickson and Gang (2002) report that this change exceeds a .01 level of significance. They reach to a similar conclusion when comparing the MSA component at Shurmai to the LSA component at Kakwa Lelash. This change occurs precisely at the time that climate is ameliorating in East Africa. The transition from embedded to direct resource procurement strategies may not be as gradual as suggested by Gang (1997:263).

There is also a distinct difference in land-use patterns indicated by the distribution of MSA and LSA sites. The emergence of this new land-use strategy also suggests a trend towards strategic planning or "planning depth" (Binford 1989). This concept, which is usually accompanied by curation, may signify the emergence of modern human thought in the Late Pleistocene, a hypothesis tentatively supported by analysis of the stone tool assemblages from Shurmai and Kakwa Lelash Rockshelters (Dickson and Gang 2002). Settlement pattern data contained herein provide supporting evidence that a dramatic change occurred.

These changes in technology and settlement pattern may, however, reflect an intensification of land use by the LSA occupants of the Mukogodo Hills due to increased internal population growth or to encroachment by other food-producing cultures. The shift to the intensified LSA mode probably preceded the introduction of pastoralism into the area, and it is logical to hypothesize that population growth was a contributing factor.

Thus, null hypothesis cannot be rejected because even though there is some circumstantial evidence for a correlation between gross climatic change and cultural change, other alternate hypotheses cannot be ruled out, nor have they been adequately tested in the current study.

A BRIEF HISTORY OF THE MUKOGODO HILLS REGION BASED ON ARCHAEOLOGICAL EVIDENCE

Foraging is the principal way that all people have made their living from the inception of our species through the early parts of the Holocene. East Africa has the longest archaeological record of human ancestry found anywhere in the world. Australopithecines, the oldest group of hominids, are divided into a number of species that are bipedal (capable of walking upright on two legs). The earliest example, *Australopithecus.anamensis*, comes from the Lake Turkana region of Kenya and is dated to between 3.9 and 4.2 million years ago (Leakey et al. 1995). Australopithecines are found both north in Ethiopia and south in Tanzania along Africa's great Rift Valley.

The Rift Valley is a remarkable geological feature. It extends over 4,830 km across most of East Africa where it is occupied by Lakes Turkana, Tanganyika, Kivu, and Olduvai Gorge, and into southwestern Asia where it forms the Red Sea, the Dead Sea, and the Jordan Valley. Within Africa, this valley is a volcanically active region. Millions of years of ash are layered into the thick geological strata where many hominid fossils are discovered. One of the largest volcanic mountains flanking the Rift Valley is Mt. Kenya, which at 5,199 m is the tallest mountain in Kenya, and the second tallest in Africa. Its snow-capped peak stands a few hundred km east of the Rift Valley, and about 50 km south of the Mukogodo Hills.

There is no archaeological evidence of any hominid occupation outside the Rift Valley for many millions of years. Mt. Kenya remained volcanically active into the Pleistocene, which began 2 million years ago. The basalt and lava fields created during these ancient eruptions may, in fact, be the source of many of the high quality nonlocal raw lithic materials that would have been used by residents of the Mukogodo Hills and Ewaso Ng'iro Plains two million years later. The sub-basement metamorphic rocks that underlie the Mukogodo Hills are Precambrian in age and are overlain by younger sediments. On the slopes of Mt. Kenya, volcanic material overlays the metamorphic complex. It is conceivable that Australopithecine, or even early Homo fossils, could be forthcoming from these sediments, but it seems unlikely. There are no known fossil mammal-bearing deposits in the region.

Compared to the Rift Valley, occupation of the Mukogodo Hills seems to occur very late. Bypassed by the complete line of *Homo* species, such as *H. habilis* (the first confirmed tool-maker in the world at 2.2 million years ago) and *H. erectus* (the first hominid to travel outside of Africa), the Mukogodo Hills are first inhabited by *H. sapiens* – fully anatomically modern human beings. Of course, it is possible and even likely that the Mukogodo Hills region was visited and perhaps even occupied by earlier Pleistocene hominids, but evidence of their presence is not preserved. Indeed, if *H. erectus* populations did not inhabit the region, it would be a major surprise. That ubiquitous hominid who left Africa at least 1.7 million years ago must certainly have left his mark on this tiny spot in East Africa (Gabunia et al. 2000).

Yet, Early Stone Age tools that would have accompanied *H. erectus* or early *H. sapiens* are completely absent from the Mukogodo Hills archaeological record. Perhaps future geoarchaeological investigations will reveal that this absence is merely the result of natural processes that have buried or destroyed any early evidence for *Homo* sp. in the region. For now we must be content to conclude that by 45,000 B.P., human foragers had reached Shurmai Rockshelter in the Mukogodo Hills.

At that time, the study region was significantly cooler than at present. The Liki Glaciation on Mt. Kenya was advancing and many East African lakes were at extremely low levels or even dried up completely dry. Vegetation belts were pushed downward on mountainsides. It is possible that during this cooler period, the Mukogodo Forest was replaced by montane species.

Occupants of the region tended to make expedient tools from locally available raw materials. The flakes upon which these tools were made are usually fairly large, with thick platforms. Archaeological site types associated with these Middle Stone Age people are not diverse and are fairly evenly distributed across the landscape. Considered together these data indicate that the manufacture of stone tools was incidental to food collection. Though Shurmai Rockshelter is a unique site because it is a rockshelter, the MSA assemblage there is unremarkable in the sense that it differs little in composition from the large open-air sites such as that represented at Kipsing.

If the Kipsing site (GnJm 23) is as old the Shurmai Rockshelter site (about 45,000 years ago), we can expect that it was occupied for a reason similar to that of later pastoralists and the British after them — water. A few kilometers west of the Kipsing site, the Kipsing River passes through a natural granitic dike, that channels it toward the shallow Pre-Cambrian bedrock at Kipsing. This means that the Kipsing River has been constrained to this spot for countless millennia. Even during times of peak aridity, groundwater was probably available here. Furthermore, during the aridity of the Late Pleistocene glacial epoch, glacial runoff may have surpassed rainwater as a more dependable seasonal water source. The Kipsing River, in part, is a product of glacial runoff from Mt. Kenya. The Middle Stone Age peoples appear to have carried on a range of activities at sites like Kipsing, Shurmai, and other sites, perhaps opportunistically exploiting specific parts of the environment and moving on to new locations when the current place was depleted.

During much of the Late Pleistocene, a gradual modification of the Stone Age toolkit takes place. Notably, lithic artifacts become smaller and there is increasing reliance upon nonlocal stone as the raw material for lithic manufacture. This progress is linear and gradual until about 14,000 years ago when the process accelerates. In addition to dealing with a rapidly and dramatically warming climate, terminal Pleistocene foragers undoubtedly faced a changing ecosystem. Over the next 10,000 years, there is substantial evidence that water supplies became more abundant, and ultimately, a savanna/grassland ecosystem began to evolve. If primary productivity increased and foragers were able to take advantage of the new ecology, then they may have experienced their own population spike.

Whatever the reason—population growth, ecological change, or even biological change—foragers of the Late Pleistocene began to use the landscape quite differently than their predecessors. Rather than employing an opportunistic strategy, they began to deploy themselves logistically. The archaeological record shows that Later Stone Age sites like Kakwa Lelash are distinctly different from the other small scattered LSA sites found primarily on the ecotone boundary between the uplands and lowlands. This area was completely populated by foragers prior to pastoralist intervention in the mid Holocene. Faunal assemblages found in the upper layers of Shurmai Rockshelter show extensive evidence of this interaction. It is likely that the pastoralist adaptation would have first encompassed the savanna environment and later progressed into the uplands. Shurmai lies in the lowlands at the very edge of the savanna environment. The Mukogodo are known to have inhabited caves at least 20-30 km into the upland forest region, at elevations 300-500 m higher than the savanna floor. The area surrounding these rockshelters would have been suitable for a foraging adaptation for much longer. Additionally, these people are either expending greater effort to procure nonlocal raw materials, or they are increasing their seasonal nomadic range to incorporate areas where nonlocal materials are present. In either case, the Pleistocene/Holocene transition witnessed major land-use intensification by foragers.

Changing climatic conditions are significant because of ecological changes that accompany shifts in temperature and rainfall patterns. Palynological evidence shows that clusters of plant species which, taken together, make up a local vegetation pattern will expand, shrink, shift elevation, or otherwise change through time. Less is known about animal responses to these changes in prehistoric times, though we know from modern studies that environmental shifts have dramatic effects upon local wildlife. Foragers in Kenya, as food collectors and rather than food producers, are dependent upon the ecosystem of which they are a part. The assumption of most researchers is that prior to the introduction of food-producing systems, human populations were small, scattered, and had very little impact on ecological change. Thus, most models of prehistoric human adaptations start with a review of climatic conditions and presume that temperature and rainfall patterns affected human use of the landscape. Humans are seen as predators existing in a state of equilibrium with the remainder of the ecosystem, only changing their land-use and subsistence patterns as needed due to larger changes in the ecosystem. In order to accommodate growing populations, African foragers would either need to find a way of producing food, or they would need to become more efficient hunters and gatherers.

One way that populations of foragers might have supported their growing numbers would have been to find a reasonably reliable and relatively stationary food source that would allow some form of semi-permanent settlement. This in itself was critical to the success of some other complex foraging societies that developed around the world, such as the Ainu and the Kwakiutl of the Pacific Rim. While wild food resources in Kenya are abundant, none are both stationary and reproduce quickly enough to permit the level of exploitation required to support a large sedentary population with a complex sociopolitical system.

If stone tool technology is any indication, prehistoric African hunters undoubtedly intensified their use of the landscape. The continuous refinement and appearance of specialized projectiles and bone implements presumably coincided with more efficient hunting strategies. The result of this would have been more successful adaptation and a consequent a rise in population. These specialized toolkits began to appear in Kenya about 30 to 40 thousand years ago, marking the transition from the African Middle Stone Age to the Late Stone Age (Ambrose 1998).

The arrival of pastoralism in Kenya is comparatively late, only within the last 4,000 to 5,000 years. It is important to note that no evidence currently exists regarding domestication of any animal species by the indigenously people of East Africa. It is also significant that only cattle pastoralists were successful enough to sweep across the African continent, replacing the foraging mode of subsistence in most of the places they visited. And finally, it was not until these pastoralists were iron-wielding that they were able to penetrate the habitats other than the savanna.

Yet, if East Africa is so well suited to pastoralism, why wasn't pastoralism a local innovation? Also, what is it about iron that enabled pastoralists to make their adaptation to East Africa complete? The answers to each question are important for understanding African history.

Chapter 5: Summary and Conclusions

However, understanding the ecology of pastoralism and foraging can only constitute one part of the study of prehistory. As pointed out by Ingold (1980), foragers and pastoralists may have decidedly different worldviews that are essentially incompatible. Changing from hunting to herding requires more than a modification in subsistence; it also requires a dramatic shift in ideology and social structure. It is easier, and perhaps more logical, that hunters would modify or intensify their use of the landscape rather than develop a new mode of production. This trend toward conservatism means that societies are expected to make difficult transitions in subsistence only as a later means of resort.

Eventually, however, if population growth is unchecked, it may be more feasible to become food producers than to continue to intensify hunting and gathering practices. This apparently did happen in Kenya with certain foods, notably finger millet (Harlan 1992). Some groups that lived in areas suitable for agriculture were able to make this transition effectively, but peoples living in the dry savanna regions did not develop extensive agriculture. Even today, agriculture is extremely limited in the dry savanna regions, including the Mukogodo Hills.

The process by which the first domestication of cattle occurred is unimportant to our current discussion, but its impact on Sub-Saharan Africa is quite significant. Cattle pastoralism was introduced to, or perhaps domesticated in, North Africa as early as 7,500 years ago, spreading throughout much of the Saharan region during times that were wetter than today. The spread of pastoralism into sub-Saharan Africa was hindered by the intertropical convergence zone beneath the Sahel, and the concomitant presence of the tsetse fly. However, as conditions became more arid after 4,000 B.P. and such barriers to southern expansion were removed, particularly tsetse fly habitat, pastoralism spread rapidly into eastern Africa (Smith 1992:72).

It is unknown why indigenous foragers did not domesticate any of the numerous game species in East Africa. The answer is most likely related to one of two possibilities: either none of the species in East Africa are suitable for domestication, or there was not significant pressure on the foragers to attempt to domesticate a wild species. Either option could be true. Even today, none of the many herding herbivores have been successfully domesticated. Nonetheless, once introduced, cattle pastoralism was a highly successful subsistence pattern for the dry savanna environment. Herders can produce more from a single acre of grassland more productive than can foragers. When competing for the same grassland resources, foragers find themselves at a selective disadvantage (Cashdan 1984:327). More efficient land use converts directly to higher reproductive fitness and higher population densities for pastoralists. Foragers are forced to retreat to more marginal areas – areas that are desired by neither pastoralists nor horticulturalists. Given enough time, there are probably very few places on the planet that are desirable by neither herders nor farmers. In the millennia following the introduction of food production in East Africa, foragers were undoubtedly pushed into the most marginal of habitats, including the upland areas of the Mukogodo Hills.

Once pastoralism was introduced and became successful, the change is irreversible—barring some dramatic event—for the following reasons. First, population density rises significantly to the point where foraging becomes impossible because the carrying capacity of the ecosystem for a foraging adaptation is exceeded. As noted previously, foraging typically requires a social environment where mobility and usufruct rights are unrestricted. Even if some pastoralists wanted to return to foraging, competition with other pastoralists for the same resources would make this difficult.

Second, the point where pastoralism enters an ecosystem is a threshold event. It represents the time at when humans began to have a significant impact upon the environment. As previously mentioned, large-scale pastoralism permanently changes the proportions of game animals available in the environment. While these changes make pastoralism productive, they work against foragers.

Once the use of iron predominates in the Mukogodo Hills, the archaeological record becomes less visible. Stone tools and their abundant debitage disappear. This most important archaeological marker is replaced by ceramics. This is no coincidence, since control of fire is necessary for both the production of pottery and of iron. Iron Age sites are broadly scattered throughout the study region, similar to the distribution of modern pastoralist sites. Because this settlement pattern is notably distinct from the Later Stone Age sites, it seems plausible that pastoralists arrived on the scene with both iron and pottery in their toolkit. If so, then certainly pastoralists arrived only within the last 2,000 years. This makes it all the more likely that they arrived when grass was plentiful on the plain.

During the Iron Age, foragers may have abandoned the plains below for exclusive occupation of the upland habitats. Initially, the uplands were probably less desirable to the pastoralists, but the ability of pastoralism to support larger numbers of people undoubtedly resulted in continuous pressure for them to expand their range. This resulted in the circumscription of foragers by the more populous pastoral groups.

Cronk (1989b) has specifically shown that the transition of the Mukogodo foragers to pastoralists occurred as a direct result of the desire to intermarry between groups. In order to gain Maasai wives, the foragers had to acquire stock as bridewealth. This meant joining the pastoralist economy upon at least a part-time basis. Ultimately, this transformation became complete, and now there is almost no indication that the Mukogodo Maasai were once foragers. Interestingly, a similar process may be impacting the Maasai pastoralists now. Rather than bridewealth, Maasai men can now find wage labor and establish their own households without working through traditional Maasai channels. Only time will tell if this change will eventually lead to the assimilation of traditional Maasai culture with other Kenyans who have adopted a market economy.

The archaeology of the Mukogodo Hills does not end with the abandonment of rockshelters and the full transition to pastoralism. We enter a period of historic archaeology – a time of lorries and guns. The last fifty years have seen the prehistoric watering hole at Kipsing become an important center of activity. Most of the buildings built by the British at Kipsing during under colonial administration are gone, but traces of the British legacy remain. A bore-hole at Kipsing is still used by the Kenya government to provide water to peacekeeping military forces that have camped nearby since the warfare in 1996. The British also built a dam across the Kipsing River, which now appears as a wall crossing the River, for it has become completely "sanded-in" behind. The British are gone, but the pastoralists still take their water here, sometimes sitting in the shade near the abandoned dam.

The dominant pastoralist group in the area are the Samburu. This is their home, and most of the terminology for geographic features are taken from their native language. They maintain very large herds of cattle, goats, sheep, and camel, and seasonally migrate very long distances. The Samburu visit Kipsing mainly to trade at the Kipsing Food Security Store, a small shop sponsored by the Catholic Kipsing Academy.

Someone with a careful eye will find numerous artifacts and features at Kipsing that extend from the present back to the Middle Stone Age. In places you might find a scrap of plastic, a thick basalt knife, a chert flake, and a stone cairn within 10 ft of each other. Kipsing today is the home of a twenty or so Turkana families. Most of them have migrated from Baragoi to the north, or from Isiolo to the east around 1993. They are very poor by pastoralist standards, but graze what livestock they have with the consent of the Samburu. Unbeknownst to them, they are living on a site that has been inhabited, perhaps intermittently, for at least the past 45,000 years. I asked one elderly Turkana man why he wanted to stay at Kipsing. He replied, "I like this place, because even if you have only one or two goats you can survive. Our main problem is only the raiders who come and steal our livestock, leaving us empty handed… I do love staying at Kipsing. We are not going back to our homeland where we used to stay… We are just going to stay here until we die." (M. Emenai, translated taped interview, August 2000).

SUGGESTIONS FOR FURTHER RESEARCH

Archaeological investigations in the Mukogodo Hills region of Central Kenya, while no longer in their infancy, are far from complete. To date, excavations at Shurmai and Kakwa Lelash rockshelters have given us an idea of the time depth of archaeological deposits in the area, and have also provided a model of lithic technology and behavioral change though time (Dickson and Gang 2002; Gang 2001; Kuehn and Dickson 1999). This volume adds to that research by providing an alluvial chronology to assist in temporally placing archaeological sites discovered in buried contexts, and by providing a predictive site model that takes into account the extent to which archaeological site distribution and preservation has been affected by ongoing natural site formation processes. Additionally, site survey has provided a model of land-use change through time. Clearly, we now have a solid foundation upon which to structure future research.

It should now be clear that the Mukogodo Hills-Ewaso Ng'iro Plains are likely to yield important information about two key periods in East African prehistory. The first is the transitional period between the Middle and Later Stone Age, and the second is the transition from the Later Stone Age to the Neolithic (i.e. the adoption of pastoralism). These important avenues of research are also being pursued by other East African archaeologists (Ambrose 1998; Dickson and Gang 2002; Gifford-Gonzalez 1998; Gifford-Gonzalez 2000; Marean 1992; Mutundu 1999; Yellen et al. 1995). The hypotheses tested in this dissertation, and those presented by Dickson and Gang (2002) for this area, need to be tested against a larger, independent data set. In order to make a significant contribution in these areas, the following research suggestions are offered:

First, expand the survey to the plains with a research design specifically intended to test the land-use model presented in Chapter 4: Site Survey. The surveys we conducted between 1996 and 2000 were tremendous archaeological enterprises and monumental feats of logistical organization. Results of those surveys are the foundation upon which we should improve and base future survey efforts. In order to test whether the land-use patterns observed in the Mukogodo foothills holds true, a larger continuous area should be surveyed. Ideally this area would be close to the current study area, but not overlapping it, thus providing a measure of independence. An ideal location would be southeast of Kipsing in the adjacent Endare Ongare valley (also a tributary of the Kipsing River). That valley was also used by the Mukogodo foragers and has never been visited by archaeologists.

Whether or not the survey coverage is expanded, ultimately special attention must be paid to the stratified sites discovered during survey. This survey discovered two buried sites, GnJm 38 and 27, but only the Lenkiteng site appeared to have much depth. The Lenkiteng site was test excavated in 2000 and is currently undergoing analyses.

One objective of expanding survey coverage is to discover additional rockshelter sites. The Mukogodo Hills have been extensively surveyed, and the two largest rockshelters have now been investigated. While both rockshelters have unexcavated sediments, it is unclear whether or not excavating them would yield independent data that could either test the hypotheses presented herein and in Dickson and Gang (2002), or add to our knowledge about occupants of the region. Rockshelters discovered elsewhere, for example in the Endare Ongare valley, would definitely provide us with independent data with which to test these hypotheses.

Perhaps the most important thing we can do in the future is to refine the chronologies presented herein. The only way to remedy this is to find additional stratified sites and date the material found therein. Refining the alluvial chronology is less problematic since natural exposures are ubiquitous. Simply, additional time and money must be expended to collect and analyze the data.

Finally, since the timing and nature of the transition to pastoralism is one of most significant issues we can address that is also relevant to Kenyan archaeology in general, it is incumbent that we become experts in pastoralist archaeology. One of the fundamental problems of archaeologists investigating middle Holocene-aged sites in East Africa is determining whether sites were created by foraging or pastoralist groups (Gifford-Gonzalez 1998; Gifford-Gonzalez and Kimengich 1984; Marean 1992; Marshall and Stewart 1995). I suggest that a modest program in pastoralist ethnoarchaeology would enable us to better understand the range of variability that we are likely to see in the pastoralist archaeological record.

REFERENCES CITED

Ambrose, S. H.
    1984   The Introduction of Pastoral Adaptations to the Highlands of East Africa. In *From Hunters to Farmers*, edited by J. D. Clark and S. Brandt, pp. 212–239. University of California Press, Berkeley.

    1998   Chronology of the Later Stone Age and Food Production in East Africa. *Journal of Archaeological Science* 25(4):377–392.

    2001   Paleolithic Technology and Human Evolution. *Science* 291:1748–1753.

Baker, R.
    1975   "Development" and the Pastoral Peoples of Karimoja, North-East Uganda: An Example of Treatment of Systems. In *Pastoralism in Tropical Africa*, edited by T. Monod, pp. 187–205. Oxford University Press, Oxford.

Barthelme, J. W.
    1985   *Fisher-Hunters and Neolithic Pastoralists of East Turkana, Kenya.* Oxford University Press, London.

Bergonzini, L., C. Francoise and G. Francoise
    1997   Paleoevaporation and Paleotranspiration in the Tanganyika Basin at 18,000 Years B.P. Inferred from Hydrologic and Vegetation Proxies. *Quaternary Research* 47:295–305.

Binford, L. R.
    1979   Organization and Formation Processes: Looking at Curated Technologies. *Journal of Anthropological Research* 35:255–273.

    1980   Willow Smoke and Dogs' Tails: Hunter-Gatherer Settlement Systems and Archaeological Site Formation. *American Antiquity* 45(1):4–20.

    1989   Isolating the Transition to Cultural Adaptations: An Organizational Approach. In *The Emergence of Modern Humans: Biocultural Adaptations in the Later Pleistocene*, edited by E. Trinkaus, pp. 18–42. Cambridge University Press, Cambridge.

Boone, J. L. and E. A. Smith
    1998   Is It Evolution Yet? A Critique of Evolutionary Archaeology. *Current Anthropology* 39:S141–S173.

Bourliere, F.
    1965   Densities and Biomasses of Some Ungulate Populations in Eastern Congo and Ruanda, with Notes on Population Structure and Lion/Ungulate Ratios. *Zoologica Africana* 1:199–207.

Bower, J.
    1986   A Survey of Surveys: Aspects of Surface Archaeology in Sub-Saharan Africa. *African Archaeological Review* 4:21–40.

    1991   The Pastoral Neolithic of East Africa. *Journal of World Prehistory* 5:49–82.

Bower, J., C. M. Nelson, A. F. Waibel and S. Wandibba
    1977   The University of Massachusetts' Later Stone Age/Pastoral Neolithic Comparative Study in Central Kenya: An Overview. *Azania* 12:119–146.

Brooks, A. S.
    1996   Behavior and Human Evolution. In *Contemporary Issues in Human Evolution*, edited by W. E. Meickle, F. C. Howell and N. G. Jablonski, pp. 135–166. California Academy of Sciences, San Francisco.

Brooks, A. S., D. M. Helgren, J. S. Cramer, A. Franklin, W. Hornyak, J. M. Keating, R. G. Klein, W. J. Rink, H. Schwarcz, J. N. L. Smith, K. Stewart, N. E. Todd, J. Verniers and J. E. Yellen
    1995   Dating and Context of 3 Middle Stone-age Sites with Bone Points in the Upper Semliki Valley, Zaire. *Science* 5210:548–553.

Butzer, K. W., G. L. Isaac, J. L. Richardson and C. Washbourne-Kamau
    1972   Radiocarbon Dating of East African Lake Levels. *Science* 175:1069–1076.

Campbell, B. C.
    1995   *Human Ecology.* 2nd ed. Aldine de Gruyter, New York.

Cann, R. L., M. Stoneking and A. C. Wilson
    1987   Mitochondrial DNA and Human Evolution. *Nature* 329:111–112.

Cashdan, E. A.
    1984   The Effects of Food Production on Mobility in the Central Kalahari. In *From Hunters to Herders: The Causes and Consequences of Fod Production*, edited by J. D. Clarke and S. Brandt, pp. 311–327. University of California Press, Berkeley.

Cassanova, J. and C. Hillaire-Marcel
    1992   Chronology and Paleohydrology of Late Quaternary High Lake Levels in the Manyara Basin (Tanzania) from Isotopic Data (^{18}O, ^{13}C, ^{14}C, Th/U) on Fossil Stromatolites. *Quaternary Research* 38:205–226.

Castro, A. P.
    1993   Kikuyu Agroforestry: An Historical Analysis. *Agriculture, Ecosystems and Environment* 46:45–54.

References

Chagnon, N. A.
1997 *Yanomamö*. 5th ed. Harcourt Brace College Publishers, Orlando, Florida.

Clark, J. D.
1970 *The Prehistory of Africa*. Praeger, New York.

Clark, J. D. and M. R. Kleindienst
1974 The Stone Age Cultural Sequence. In *Kalambo Falls Prehistoric Site*, edited by J. D. Clark, pp. 71–106. Cambridge University Press, Cambridge.

Coetzee, J. A.
1964 Evidence for a Considerable Depression of the Vegetation Belts During the Upper Pleistocene on the East African Mountains. *Nature* 204(4958):564–566.

1967 Pollen Analytical Studies in East and Southern Africa. In *Paleoecology of Africa and of the Surrounding Islands and Antarctica*, edited by Z. Bakker and E. v. Meine. vol. 3, pp. 1–17. Balkema, Cape Town.

Cronk, L.
1989a *The Behavioral Ecology of Change Among the Mukogodo of Kenya*. Unpublished Ph.D. Dissertation, Northwestern University, Evanston, Illinois.

1989b From Hunters to Herders: Subsistence Change as a Reproductive Strategy among the Mukogodo. *Current Anthropology* 30:224–234.

2000 From 'True Dorobo' to 'Mukogodo Maasai': Contested Ethnicity and Cultural Change in Kenya. Unpublished Manuscript. Department of Anthropology, Rutgers University, New Brunswick, New Jersey.

Dibble, H. L. and O. Bar-Yosef
1995 *The Definition and Interpretation of Levallois Technology*. Prehistory Press, Madison, Wisconsin.

Dickson, D. B.
1993 An Ethnoarchaeological Site Survey of Dry Rockshelters Formerly Occupied by the Mukogodo Maasai Hunter-Gatherers of Eastern Laikipia District, Kenya, East Africa. Paper presented at the The 58th Annual Meeting of The Society for American Archaeology, St. Louis, Missouri.

Dickson, D. B. and G.-Y. Gang
2002 Evidence of the Emergence of "Modern" Behavior in the Middle and Later Stone Age Lithic Assemblages at Shurmai Rockshelter (GnJm1) and Kakwa Lelash Rockshelter (GnJm2) in the Mukogodo Hills of North-Central Kenya. *African Archaeological Review* 19:1–26.

Dickson, D. B. and D. Kuehn
1997 A Preliminary Outline of Late Quaternary Paleoenvironment and Human Occupation of the Mukogodo Hills of Southern Isiolo and Northeastern Laikipia District, Kenya. Paper presented at the 62nd Annual Meeting of the Society for American Archaeology, Nashville, Tennessee.

Dickson, D. B., D. Kuehn and G.-Y. Gang
1998 The Middle and Later Stone Age Lithic Assemblages at Shurmai Rockshelter (GnJm1) in the Mukogodo Hills of North-Central Kenya. Paper presented at the The 63rd Annual Meeting of The Society for American Archaeology, Seattle, Washington.

Dodd, J. L.
1994 Desertification and Degradation in Sub-Saharan Africa. *Bioscience* 44:28–34.

Duke, J. A.
1984 Handbook of Energy Crops. Manuscript on file, Center for New Crops and Plant Products, Purdue University, West Lafayette, Indiana.

Ehret, C.
1995 Yaakuans as Long-term Actors in the Culture History of Northern Kenya. Paper presented at the 94th Annual Meeting of the American Anthropological Association, Washington, D.C.

Eliot, C.
1905 *The East Africa Protectorate*. Frank Cass & Co. Ltd., London.

Ellis, J. E. and D. M. Swift
1988 Stability of African Pastoralist Ecosystems: Alternate Paradigms and Implications for Development. *Journal of Range Management* 41:450–459.

Emiliani, C.
1955 Pleistocene Temperatures. *Journal of Geology* 63:538–578.

English, J., M. Tiffen and M. Mortimore
1994 *Land Resource Management in Machakos District, Kenya, 1930–1990*. Vol. 5. World Bank, Washington, D.C.

Evans-Pritchard, E. E.
1940 *The Nuer*. Oxford University Press, Oxford.

Farrand, W. R., R. W. Redding, M. H. Wolpoff and H. T. Wright
1976 *An Archaeological Investigation on the Loboi Plain, Baringo District Kenya*. Museum of

Anthropology, University of Michigan, Ann Arbor.

Foley, R. A.
1980 Spatial Component of Archaeological Data: Off-site Methods and Some Preliminary Results from the Amboseli Basin, Southern Kenya. In *Proceedings of the 8th Panafrican Congress of Prehistory and Quaternary Studies*, edited by R. E. Leakey and B. A. Ogot, pp. 39–40. TILLMIAP, Nairobi.

1981 *Off-site Archaeology and Human Adaptation in East Africa: An Analysis of Regional Artefact Density in the Amboseli, Southern Kenya*. Cambridge Monographs in African Archaeology 3 BAR International Series 97, Oxford.

Fosbrooke, H. A.
1957 Prehistoric Wells, Rainponds and Associated Burials in Northern Tanganyika. In *Third Panafrican Congress on Prehistory, Livingstone, 1955*, edited by J. D. Clarke, pp. 326–255, Cambridge University Press, Cambridge.

Frayer, D. W., M. H. Wolpoff, A. G. Thorne, F. H. Smith and G. G. Pope
1993 Theories of Modern Human Origins: The Paleontological Test. *American Anthropologist* 96:424–428.

Gabunia, L., A. Vekua, D. Lordkipanidze, C. C. Swisher III, R. Ferring, A. Justus, M. Nioradze, M. Tvalchrelidze, S. C. Antón, G. Bosinski, O. Jöris, M. A. de Lumley, G. Majsuradze and A. Mouskhelishvili
2000 Earliest Pleistocene Hominid Cranial Remains from Dmanisi, Republic of Georgia: Taxonomy, Geological Setting, and Age. *Science* 288(5468):1019–1025.

Gang, G.-Y.
2001 Middle and Later Stone Ages in Mukogodo Hills, Central Kenya. Cambridge Monographs in African Archaeology 52, Bar International Series 954. BAR Publishing, Oxford, England.

1997 *Comparative Analysis of Lithic Materials Recovered from Shurmai (GnJm 1) and Kakwa Lelash (GnJm 2) Rockshelters, Kenya*. Unpublished Ph.D. Dissertation, Texas A&M University.

Gifford-Gonzalez, D.
1998 Early Pastoralists in East Africa: Ecological and Social Dimensions. *Journal of Anthropological Archaeology* 17(2):166–200.

2000 Animal Disease Challenges to the Emergence of Pastoralism in Sub-Saharan Africa. *African Archaeological Review* 17:95–139.

Gifford-Gonzalez, D. and J. Kimengich
1984 Faunal Evidence for Early Stock-Keeping in the Central Rift of Kenya: Preliminary Findings. In *Origins and Early Development of Food-Producing Cultures in Northeast Africa*, edited by L. Krzyzaniak, pp. 457–471. Polish Academy of Sciences, Poznan.

Gile, L. M., F. F. Peterson and R. B. Grossman
1966 Morphological and Genetic Sequences of Carbonate Accumulation in the Desert Soils. *Soil Science* 101:346–360.

Goodwin, A. J. H. and L. van Riet
1929 The Stone Age Cultures of South Africa. *Annals of the South African Museum* 27:1–289.

Haberyan, K. A. and R. E. Hecky
1987 The Late Pleistocene and Holocene Stratgraphy and Paleolimnology of Lakes Kivu and Tanganyika. *Palaeogeography, Palaeoclimatology, Palaeoecology* 61:169–197.

Hackman, B. D., T. J. Charsley, J. Kagasi, R. M. Key, W. S. Siambi and A. F. Wilkinson
1989 *Geology of the Isiolo Area*. Republic of Kenya Ministry of Environment and Natural Resources, Mines and Geology Department, Nairobi.

Hamilton, A. C.
1982 *Environmental History of East Africa: A Study of the Quaternary*. Academic Press, New York.

Hardin, G.
1968 The Tragedy of the Commons. *Science* 162:1243–1248.

Hare, P. E., G. A. Goodfriend, A. S. Brooks, J. E. Kokis and D. W. von Endt
1993 Chemical Clocks and Thermometers: Diagenetic Reactions of Amino Acids in Fossils. *Carnegie Institute of Washington Yearbook* 92:80–85.

Harlan, J. R.
1992 Indigenous African Agriculture. In *The Origins of Agriculture*, edited by C. W. Cowan and P. J. Watson, pp. 59–70. Smithsonian Institution Press, Washington, D.C.

Hays, J. D., J. Imbrie and N. J. Shackleton
1976 Variations in the Earth's Orbit: Pacemaker of the Ice Ages. *Science* 194:1121–1132.

Holmgren, P., E. J. Masakha and H. Sjoeholm
1994 Not All African Land Is Being Degraded: A Recent Survey of Trees on Farms in Kenya Reveals Rapidly Increasing Forest Resources. *Ambio Stockholm* 23(7):390–395.

Huxley, E.
1935 *White Man's Country: Lord Delamere and the Making of Kenya Volume 2*. Chatto and Windus, London.

References

Imbrie, J., J. D. Hays, D. G. Martinson, A. McIntyre, A. C. Mix, J. J. Morley, N. G. Pisais, W. L. Prell and N. J. Shackelton
- 1984 The Orbital Theory of Pleistocene Climate: Support for a Revised Chronology of the Marine $\delta^{18}O$ Record. In *Milankovitch and Climate, Part I*, edited by A. L. Berger, pp. 269–305. Reidel, Amsterdam, The Netherlands.

Ingold, T.
- 1980 *Hunters, Pastoralists and Ranchers*. Cambridge University Press, Cambridge.

Jacobs, A. H.
- 1965 African Pastoralists: Some General Remarks. *Anthropological Quarterly* 38:144–154.

Karega-Munene
- 1996 The East African Neolithic: An Alternative View. *African Archaeological Review* 13:247–254.

Klein, R. G.
- 1984 Mammalian Extinctions and Stone Age People in Africa. In *Quaternary Extinctions: A Prehistoric Revolution*, edited by P. S. Martin and R. G. Klein, pp. 553–573. University of Arizona Press, Tucson.
- 1999 *The Human Career: Human Biological and Cultural Origins*. University of Chicago Press, Chicago.

Kuehn, D.
- 1999 Preliminary Geomorphology and Alluvial History of the Tol and Kipsing Rivers of Kenya, East Africa. Manuscript on file, Department of Anthropology, Texas A&M University, College Station.

Kuehn, D. and D. B. Dickson
- 1999 Stratigraphy and Non-Cultural Site Formation at the Shurmai Rockshelter (GnJm1) in the Mukogodo Hills of North-Central Kenya. *Geoarchaeology* 14:63–85.

Kuehn, D. D., S. A. Jennings, D. B. Dickson and C. S. Caran
- 1996 *Late Pleistocene and Holocene Climatic Variability in the Tol River Area, Upper Ewaso Ng'iro Drainage Basin, North-Central Kenya*. Department of Anthropology, Texas A&M University.

Kusimba, S. B.
- 1999 Hunter-gatherer Land Use Patterns in Later Stone Age East Africa. *Journal of Anthropological Archaeology* 18(2):165–200.

Kyule, M. D., S. H. Ambrose, M. P. Noll and J. L. Atkinson
- 1997 Pliocene and Pleistocene Sites in Southern Narok District, Southwest Kenya. *Journal of Human Evolution* 32(4):A9–10.

Leakey, L. B.
- 1931 *Stone Age Cultures of Kenya Colony*, Cambridge University Press, Cambridge.

Leakey, M. G., C. S. Feibel, I. McDougall and A. C. Walker
- 1995 New Four-Million-Year-Old Species from Kanapoi and Allia Bay, Kenya. *Nature* 376:565–571.

Lind, E. M. and M. E. S. Morrisson
- 1974 *East African Vegetation*. Longman, London.

Livingstone, D. A.
- 1996 Historical Ecology. In *East African Ecosystems and Their Conservation*, edited by T. R. McClanahan and T. P. Young, pp. 3–17. Oxford University Press, New York.

Lyman, R. L. and M. J. O'Brien
- 1998 The Goals of Evolutionary Archaeology: History and Explanation. *Current Anthropology* 39:615–652.

Mahaney, W. C., R. W. Barendregt and W. Vortisch
- 1989 Quaternary Glaciations and Paleoclimate of Mt. Kenya, East Africa. In *Glacier Fluctuations and Climatic Change*, edited by J. Oerlemans, pp. 13–35. Kluwer Academic Publishers, Dordrecht, The Netherlands.
- 1990 *Ice on the Equator: Quaternary Geology of Mt. Kenya, East Africa*. Caxton, Sister Bay, Wisconsin.

Marean, C. W.
- 1992 Hunter to Herder: Large Mammal Remains from the Hunter-gatherer Occupation at Enkapune Ya Muto Rockshelter, Central Rift, Kenya. *African Archaeological Review* 10:65–127.

Marshall, F. B.
- 1994 Archaeological Perspectives on East African Pastoralism. In *African Pastoralist Systems: An Integrated Approach*, edited by E. Fratkin, K. A. Galvin and E. A. Roth, pp. 17–43. Lynne Reinner, London.

Marshall, F. B., J. W. Barthelme and K. Stewart
- 1984 Early Domestic Stock at Dongodien in Northern Kenya. *Azania* 19:120–127.

Marshall, G. and K. Stewart
- 1995 Hunting, Fishing and Herding Pastoralists of Western Kenya: The Fauna of Gogo Falls. *Zooarchaeologia* 7:7–27.

Martinson, D. G., N. G. Pisias, J. D. Hays, J. Imbrie, T. C. M. Jr. and N. J. Shackelton

1987 Age Dating and the Orbital Theory of the Ice Ages: Devolopment of a High Resolution 0 to 300,000-Year Chronostratigraphy. *Quaternary Research* 27:1–29.

McBrearty, S.
1988 The Sangoan-Lupemban and Middle Stone-Age Sequence at the Muguruk Site, Western Kenya. *World Archaeology* 19(3):388–420.

McBrearty, S., L. C. Bishop and J. D. Kingston
1996 Identifying the Acheulian to Middle Stone Age Transition in the Kapthurin Formation, Baringo, Kenya. *Journal of Human Evolution* 30:563–580.

McBrearty, S. and A. S. Brooks
2000 The Revolution That Wasn't: A New Interpretation of the Origin of Modern Human Behavior. *Journal of Human Evolution* 39:453–563.

McCabe, J. T.
1989 Turkana Pastoralism – A Case Against the Tragedy of the Commons. *Human Ecology* 18:81–103.

Mellars, P. A. and C. Stringer (editors)
1989 *The Human Revolution: Behavioral and Biological Perspectives on the Origins of Modern Humans*. Edinburgh University Press, Edinburgh, United Kingdom.

Merrick, H. V., F. H. Brown and M. Connelly
1990 Sources of Obsidian at Ngamuriak and Other South-Western Kenyan Sites. In *Early Pastoralists of South-Western Kenya*, edited by P. Robertshaw, pp. 173–182. British Institute in Eastern Africa, Nairobi, Kenya.

Michels, J. W.
1990 Obsidian Dating in the Lemek-Mara Region. In *Early Pastoralists of South-Western Kenya*, edited by P. Robertshaw, pp. 52–53. British Institute in Eastern Africa, Nairobi, Kenya.

Michels, J. W., I. S. T. Tsong and C. M. Nelson
1983 Obsidian Dating and East African Archaeology. *Science* 219:361–366.

Miller, G. H., P. B. Beaumont, A. J. T. Jull and B. Johnson
1993 Pleistocene Geochronology and Paleothermometry from Protein Diagenesis in Ostrich Eggshells: Implications for the Evolution of Modern Humans. In *The Origin of Modern Humans and the Impact of Chronometric Dating*, edited by M. J. Aitken, C. B. Stringer and P. A. Mellars, pp. 49–68. Princeton University Press, Princeton, New Jersey.

Mutundu, K. K.
1999 *Ethnohistoric Archaeology of the Mukogodo in North-Central Kenya: Hunter-Gatherer Subsistence and the Transition to Pastoralism in Secondary Settings*. BAR International Series 775. BAR Publishing, Oxford.

Osmaston, H.
1989 Glaciers, Glaciations, and Equilibrium Line Altitudes on Kilimanjaro. In *Quaternary and Environmental Research on East African Mountains*, pp. 5–30. Balkema, Rotterdam, The Netherlands.

Owen, R. B., J. W. Barthelme, R. W. Renaut and A. Vincens
1982 Paleolimnology and Archaeology of Holocene Deposits North-east of Lake Turkana. *Nature* 298:523–529.

Owen-Smith, R. N.
1988 *Megaherbivores: The Influence of Very Large Body Size on Ecology*. Cambridge University Press, Cambridge.

Pearl, F. B.
2001 Late Pleistocene Archaeological and Geoarchaeological Investigations in the Mukogodo Hills and Ewaso Ng'iro Plains of Central Kenya. Ph.D. dissertation, Texas A&M University. University Microfilms, Ann Arbor.

Phillipson, D. W.
1993 *African Archaeology*. 2nd ed. Cambridge University Press, Cambridge.

Pisias, N. G., D. G. Martinson, T. C. M. Jr., N. J. Shackelton, W. L. Prell, J. Hays and G. Boden
1984 High Resolution Stratigraphic Correlation of Benthic Oxygen Isotope Records Spanning the Last 300,000 Years. *Marine Geology* 56:119–136.

Plog, S., F. Plog and W. Wait
1978 Decision Making in Modern Surveys. In *Advances in Archaeological Method and Theory*, edited by M. Schiffer, pp. 383–421. vol. 1. Academic Press, New York.

Popper, K. R.
1959 [1934] *The Logic of Scientific Discovery*. Harper and Row, New York.

Posnansky, M.
1968 Cairns in the Southern Part of the Rift Valley. *Azania* 3:181–187.

Richardson, J. L. and A. E. Richardson
1972 History of an African Rift Lake and Its Climatic Implications. *Ecological Monographs* 42:499–534.

References

Roberts, R. G., R. Jones, N. A. Spooner, M. J. Head, A. S. Murray and M. A. Smith
1994 The Human Colonization of Australia: Optical Dates of 53,000 and 60,000 Bracket Human Arrival at Deaf Adder Gorge, Northern Territory. *Quaternary Science Reviews* 13:575–586.

Robertshaw, P. T.
1995 The Last 200,000 Years (or Thereabouts) in Eastern Africa: Recent Archaeological Research. *Journal of Archaeological Research* 3:55–86.

Robertshaw, P. T. and D. Collett
1983 A New Framework for the Study of Early Pastoral Communities in East Africa. *Journal of African History* 24:289–301.

Robertshaw, P. T., T. Pilgram, A. Siiriainen and F. B. Marshall
1990 Archaeological Surveys and Prehistoric Settlement Patterns. In *Early Pastoralists of South-Western Kenya*, edited by P. T. Robertshaw, pp. 36–51. British Institute in Eastern Africa, Nairobi, Kenya.

Ross, W. M.
1927 *Kenya from Within: A Short Political History*. 1st edition, new impression ed. Allen and Unwin, London.

Schiffer, M. B. and S. Wells
1982 Archaeological Surveys: Past and Future. In *Hohokam and Patayan: Prehistory of Southwestern Arizona*, edited by R. H. McGuire and M. B. Schiffer, pp. 345–383. Academic Press, New York.

Schmidt, P. R.
1975 A New Look at Interpretations of the Early Iron Age in East Africa. *History in Africa* 2:127–236.

Schumm, S. A.
1977 *The Fluvial System*. Wiley, New York.

SCS (Soil Conservation Service)
1992 *Keys to Soil Taxonomy*. Pocahontas Press, Blacksburg, Virginia.

Semaw, S., P. Renne, J. W. K. Harris, C. S. Feibel, R. L. Bernor, N. Fesseha and K. Mowbray
1997 2.5-Million-Year-Old Stone Tools from Gona, Ethiopia. *Nature* 385:333–336.

Shackleton, N. J.
1967 Oxygen Isotope Analysis and Pleistocene Temperatures Re-assessed. *Nature* 5096:15–17.

Shackleton, N. J. and D. Opdyke
1976 Oxygen-isotope and Paleomagnetic Stratigraphy of Pacific Core V28-239, Late Pliocene to Latest Pleistocene. In *Investigation of Late Quaternary Paleooceanography and Paleoclimatology*, pp. 449–464. vol. 145. Geological Society of America, Boulder, Colorado.

Sheppard, P. J. and M. R. Kleindienst
1996 Technological Change in the Earlier and Middle Stone Age of Kalambo Falls (Zambia). *African Archaeological Review* 13:171–196.

Smith, A. B.
1992 *Pastoralism in Africa: Origins and Development Ecology*. Ohio University Press, Athens.

Spencer, P.
1973 *Nomads in Alliance: Symbiosis and Growth Among the Rendille and Samburu of Kenya*. Oxford University Press, London.

Stager, J. C., B. F. Cumming and L. Meeker
1997 A High-resolution 11,400-yr Diatom Record from Lake Victoria, East Africa. *Quaternary Research* 47:81–89.

Stager, J. C. and T. C. Johnson
2000 A 12,400 C-14 Yr Offshore Diaton Record from East Central Lake Victoria, East Africa. *Journal of Paleolimnology* 23:373–383.

Stager, J. C., P. N. Reinthal and D. A. Livingstone
1986 A 25,000-Year History for Lake Victoria, East-Africa, and Some Comments on Its Significance for the Evolution of Cichlid Fishes. *Freshwater Biology* 16(1):15–19.

Steward, J. H.
1938 *Basin-Plateau Aboriginal Sociopolitical Groups*. Smithsonian Institution Bureau of American Ethnology Bulletin 120. United States Government Printing Office, Washington D.C.

Street-Perrott, F. A., D. S. Marchand, N. Roberts and S. P. Harrison
1989 *Global Lake-level Variations from 18,000 to Ø Years Ago: A Paleoclimatic Analysis*. U.S. Department of Energy, Office of Energy Research. Copies available from DOE/ER/60304-H1.

Stringer, C. B. and P. Andrew
1988 Genetic and Fossil Evidence for the Origin of Modern Humans. *Science* 239:1261–1268.

Sutton, J. E. G.
1973 *The Archaeology of the Western Highlands of Kenya*. BAR International Series, Oxford.

Swift, D. M., M. B. Coughenour and M. Atsedu

1996 Arid and Semi-arid Ecosystems. In *East African Ecosystems and Their Conservation*, edited by T. R. McClanahan and T. P. Young, pp. 243–272. Oxford University Press, New York.

Szabo, B. J., C. V. Haynes and T. A. Maxwell
1995 Age of Quaternary Pluvial Episodes Determined by Uranium-series Dating of Lacustrine Deposits of Eastern Sahara. *Palaeogeography, Palaeoclimatology, Palaeoecology* 113:227–242.

Thouless, C. R.
1994 Conflict Between Humans and Elephants on Private Land in Northern Kenya. *Oryx* 28(2):119–127.

van Grunderbeek, M.-C., E. Roche and H. Doutrelpont
1988 *Le Premiere Âge du Fer au Rwanda et au Burundi: Archéologie et Environnement*. Institut National de Recherche Scientifique, Butare, Rwanda.

Vigilant, L., M. Stoneking, H. Harpending, K. Hawkes and A. C. Wilson
1991 African Populations and the Evolution of Human Mitochondrial DNA. *Science* 253:1503–1507.

Ward, W. E. F. and L. W. White
1971 *East Africa: A Century of Change 1870–1970*. Africana Publishing Corporation, New York.

Washbourn, C. K.
1967 Lake Levels and Quaternary Climates in the Eastern Rift Valley of Kenya. *Nature* 5116:672–673.

Washbourn-Kamau, C. K.
1970 Late Quaternary Chronology of the Nakuru-Elmenteita Basin. *Nature* 5242:253–254.

1975 Late Quaternary Shorelines of Lake Naivasha, Kenya. *Azania* 10:77–92.

Waters, M. R.
2000 Alluvial Stratigraphy and Geoarchaeology in the American Southwest. *Geoarchaeology: An International Journal* 15(6):537–557.

Waters, M. R. and D. D. Kuehn
1996 The Geoarchaeology of Place: The Effect of Geological Processes on the Preservation and Interpretation of the Archaeological Record. *American Antiquity* 61(3):483–497.

Watson, P. J., S. A. LeBlanc and C. L. Redman
1971 *Explanation in Archaeology*. Columbia University Press, New York.

Wendorf, F., R. L. Laury, R. Schild, C. V. Haynes and P. D. Damon
1975 Dates for the MIddle Stone Age of East Africa. *Science* 187:740–742.

Winterhalder, B. and E. A. Smith
1992 Evolutionary Ecology and the Social Sciences. In *Evolutionary Ecology and Human Behavior*, edited by E. A. Smith and B. Winterhalder, pp. 3–24. Aldine de Gruyter, New York.

Woodburn, J.
1972 An Introduction to Hadza Ecology. In *Man the Hunter*, edited by R. B. Lee and I. Devore, pp. 49–55. Aldine de Gruyter, New York.

Yellen, J. E., A. S. Brooks, E. Cornelissen, M. J. Mehlman and K. Stewart
1995 A Middle Stone-age Worked Bone Industry from Katanda, Upper Semliki Valley, Zaire. *Science* 5210:553–556.

Young, T. P., N. Patridge and A. Macrae
1995 Long-term Glades in Acacia Bushland and Their Edge Effects in Laikipia, Kenya. *Ecological Applications* 5(1):97–108.

APPENDIX A

SITES RECORDED DURING THE 1996 SURVEY

In the following discussion, several types of information are presented. First, a general description is given of the site's location. This description includes the following information: approximate distance and direction to a nearby major topographic feature, usually a river; and the approximate distance and direction to a nearby archaeological site. Bearings, when given, are accurate to within 2°. These measurements are only given to help relocate sites or to help the reader picture their location.

Stratigraphic position is expressed in terms of the geological units described previously (i.e., Peleta, Kipsing, Shordika, Seaku, and Tol units). In the absence of natural on-site exposures, it was often impossible to determine the exact stratigraphic context of a site. If I was unable to determine the exact stratigraphic unit from the original site forms or from visits to the site, I have listed the probable units.

For each site, a selected set of information is provided on the observed artifacts, as well as a qualitative description. G-Young Gang provided the identifications and measurements of all the lithic specimens (full data reported in Pearl 2001: appendix B).

The following descriptions are based on my field notes, as well as from the field notes and site forms prepared by D. Bruce Dickson. Following this brief overview of sites recorded during the survey is a comparative analysis of sites in the region. All interpretations, conclusions, and inferences are my own.

GnJm 7: Lithic Scatter, Early Middle Stone Age

The site is located on the point of land formed by the junction of the Seaku and Tol Rivers. It was initially discovered exposed in a shallow ditch paralleling the dirt road that crosses the Seaku River. The scatter turned out to be quite extensive, covering most of the point of land formed by the junction of the two streams. The nearest other site is GnJm 12, which is about 340 m to the southeast (bearing 110°). Buried lithics were observed embedded up to 30 cm deep in Peleta alluvium, and all lithics from this site are basalt.

One hundred and nine artifacts were collected from the main part of the site. Of these, 98.17 percent (n=107) are basalt, with less than 1 percent each of obsidian (n=1) and quartz (n=1). Average flake dimensions are 47.15 mm long, 41.33 mm wide, and 19.97 mm thick. Of the 109 artifacts, 22.02 percent (n=24) are cores. The cores tend to be flaked in a radial pattern and, generally speaking, are large (about two-thirds of the specimens have a minimum thickness over 3.5 cm). Ten stone tools were collected from the surface, including four choppers, one scraper, one hand-axe, one pick, and three cobble tools. Raw material usage, flake size, and tool types indicate that this site is most likely referable to the early MSA.

GnJm 8: Stone Cairns

Seven stone cairns were found within a 200 m diameter area. The central cairn is the largest, measuring 8 m by 6.5 m. Six surrounding cairns were about 4 m in diameter. This cairn group is located on the point of land formed by the junction of the Tol and Seaku Rivers. Like GnJm 7, this site sits on the Peleta surface. There was no other cultural material observed in the area. The nearest archaeological site is GnJm 20, located 220 m to the west (bearing 262°) across the Tol River.

GnJm 9: Ceramic Scatter, Iron Age

This site consists of numerous small, thick, ceramic sherds resting on the Tol alluvium. An estimated two vessels were represented. The nearest site is GnJm 11, approximately 940 m to the west (bearing 288°). The Lenkiteng tributary is 300 m to the west.

GnJm 10: Ceramic Scatter, Iron Age

This site is a small scatter of plain ceramic sherds resting on the Tol surface. It is located adjacent to the Tol River on the east bank. The nearest archaeological site is GnJm 11, approximately 250 m due south (bearing 180°).

GnJm 11: Ceramic Scatter, Iron Age

This site is a small scatter of plain ceramic sherds resting on the Tol surface. It is located about 290 m southeast of the Tol River. The nearest archaeological site is GnJm 10, approximately 250 m due north (bearing 360°).

GnJm 12: Lithic Scatter, Middle/Later Stone Age

This site is located on the bank of the Seaku River just upstream (south) of its junction with the Tol River, on the Peleta surface. It is in close proximity to GnJm 7 and was originally thought to be a part of that site. GnJm 7 lies 340 m to the northwest (bearing 290°). A decision was made to separate the two sites because: a) material for this site was highly localized, and b) the material types and flaking types appeared to be different at the two sites.

Two hundred and fifty-nine artifacts were collected from the site. Of these, 73.36 percent (n=190) are basalt, 14.67 percent (n=38) are chert, 10.42 percent (n=27) are obsidian, and 1.54 percent (n=4) are quartz. The average flake size is 39.05 mm long, 36.32 mm wide, and 11.57 mm thick. Of the 259 artifacts, 5.79 percent (n=15) are basalt cores. All cores were flaked in a radial pattern. Though basalt outnumbers all other material types, the flakes are significantly smaller than those from other nearby MSA-type sites (c.f., GnJm 7). This site either represents a multicomponent site or a transitional site between MSA and LSA.

GnJm 13: Lithic Scatter, Middle Stone Age

This site is located on the west bank of the Seaku River on a terrace above the floodplain north of its junction with

Peleta Creek. The nearest site is GnJm 25, approximately 1.02 km to the northeast (bearing 28°). Lithics were recovered in the gullies cutting into the terrace. Some basalt flakes appeared to be eroding from the bank itself, which is Peleta alluvium.

One hundred and fifteen artifacts were collected from the site. Of these, 86.09 percent (n=99) are basalt, 13.04 percent (n=15) are chert, and 0.87 percent (n=1) are obsidian. The average flake size is 47.39 mm long, 44.02 mm wide, and 13.42 mm thick. Of the 115 artifacts, 8.70 percent (n=10) are basalt cores. All cores were flaked in a radial pattern. Four additional cores exhibited battering, possibly from use in a chopping fashion. Two end-scrapers were also in the assemblage. Raw material and flake size strongly suggests that this site dates to the MSA.

GnJm 14: Lithic Scatter, Middle Stone Age

This site is located on the west bank of the Seaku River north of Peleta Creek in a deep gully incising the Peleta alluvium overlooking the Seaku. The nearest archaeological site is GnJm 26, which is about 225 m to the northeast (bearing 45°).

Forty-six artifacts were collected from the site. Of these, 95.65 percent (n=44) are basalt, 2.17 percent (n=1) are chert, and 2.17 percent (n=1) are obsidian. The average flake size is 42.88 mm long, 42.81 mm wide, and 11.73 mm thick. Of the 46 artifacts, 8.70 percent (n=4) are basalt cores and 2.17 percent (n=1) are chert cores. Most cores were flaked in a radial pattern (one appears randomly flaked). Two scrapers were also in the assemblage, with usewear around the entire edge. Raw material, flake size, and morphology strongly suggest that this site represents the MSA as seen at Shurmai Rockshelter (GnJm 1).

GnJm 15: Iron Smelting Site, Iron Age.

This site is a small cluster of iron slag resting on the Seaku surface upstream of the junction of the Seaku River and Peleta Creek. The nearest archaeological site is GnJm 14, approximately 1.13 km to the northeast (bearing 27°).

GnJm 16: Lithic Scatter, Middle Stone Age

This site is located on the Kipsing surface on the north bank of the Kipsing River, west of its junction with the Seaku. It consists of black basalt lithic cores and flakes of a crude aspect scattered over a large area (perhaps 100 m diameter or more). The nearest archaeological site is GnJm 18, across the Kipsing River about 360 m to the southwest (bearing 192°).

Seventy-five artifacts were collected from the site. Of these, 90.54 percent (n=67) are basalt and 9.46 percent (n=7) are chert. The average flake size is 42.09 mm long, 41.28 mm wide, and 11.80 mm thick. Of the 74 artifacts, 9.33 percent (n=7) are basalt cores. All cores were flaked in a radial pattern. Two cores exhibited evidence of battering and one of scraping. One flake tool with edge-wear was also in the assemblage. Tool type, raw material, and flake size suggest that this site represents the early MSA.

GnJm 17: Stone Cairn

This site is located on the Kipsing surface just west of the junction of the Seaku and the Kipsing rivers. It is situated on a highly eroded portion of the interfluvial area. The cairn has been partially dismantled. Its dimensions are roughly oval, about 7 m by 6 m, being slightly less than 1 m high. Two smaller piles of stones, about 1 m in diameter each, lie about 20 m southeast from the main cairn. The nearest archaeological site, GnJm 22, is just across the Seaku River, approximately 175 m to the south (bearing 172°).

GnJm 18: Lithic Scatter, Multiple Component (MSA, IRA)

This site is located on a slightly elevated portion of the interfluvial area between the Kipsing and Seaku rivers, southwest of their junction, on the Kipsing surface. Lithic flakes of black basalt were predominant, but one small pot sherd was also recovered. The nearest archaeological site is GnJm 19, approximately 200 m south (bearing 192°).

Twenty-six artifacts were collected from the site. Of these, 92.31 percent (n=44) are basalt, 3.85 percent (n=1) are obsidian, and 3.85 percent (n=1) are pottery. The average flake size is 59.05 mm long, 51.20 mm wide, and 16.31 mm thick. Of the 26 artifacts, 26.92 percent (n=7) are basalt cores, each radially flaked. There were no tools in the assemblage. Raw material use and flake type are indicative of the Middle Stone Age, while the presence of pottery indicates a later age.

GnJm 19: Lithic Scatter, Later Stone Age

This site is located on a slightly elevated portion of the interfluvial area between the Kipsing and Seaku rivers, southwest of their junction, on the Kipsing surface. The nearest archaeological site is GnJm 17, approximately 180 m southwest (bearing 230°).

Thirty-two artifacts were collected from the site. Of these, 12.50 percent (n=4) are basalt, 28.13 percent (n=9) are chert, and 59.38 percent (n=19) are obsidian. The average flake size is 25.95 mm long, 24.94 mm wide, and 7.81 mm thick. Of the 32 artifacts, 3.13 percent (n=1) are basalt cores. There are no tools in the assemblage. Artifacts were produced of local and nonlocal, exotic material. Additionally, the flake size is similar to those found in the LSA components of GnJm 1 and GnJm 2.

GnJm 20: Stone Cairn

This site is located on the west side of the Tol River on the Kipsing surface. It is west of GnJm 8 (bearing 82°), another stone cairn site. The site consists of a single cairn, oval in shape, measuring approximately 5.7 m by 5 m.

GnJm 21: Stone Cairn

This site is located on the east bank of the Seaku River, just south of its junction with the Tol on the Peleta surface. Two large cairns are built around exposed bedrock which serves as their foundation. These two cairns measure 4.0 m by 4.2 m, and 3.0 m by 2.5 m. A third cairn, 4.0 m by 4.2 m, is located 100 m to the east of these two. A fourth cairn, 3.3 m by 3.6 m, is located about 50 m southeast of the main two cairns. The nearest archaeological site is GnJm 8, located about 300 m to the northwest (bearing 288°).

GnJm 22: Lithic Scatter, Middle Stone Age

This site is located on the eroded first terrace of the east bank of the Seaku River, south of its junction with the Kipsing. A nearby archaeological site is GnJm 34, located approximately 375 m to the northeast (bearing 60°). Lithic material appears to be eroding from the Peleta alluvium. Bedrock is exposed in the riverbed here and natural cisterns have formed in it.

Fifty-nine artifacts were collected from the site. Of these, 62.71 percent (n=37) are basalt, 27.12 percent (n=16) are chert, 8.47 percent (n=5) are obsidian, and 1.69 percent (n=1) are quartz. The average flake size is 33.70 mm long, 32.36 mm wide, and 10.33 mm thick. Of the 59 artifacts, 10.17 percent (n=6) are basalt cores and 3.39 percent (n=2) are chert cores. Most cores were flaked in a radial pattern (two were bidirectionally flaked only). Seven core-tools and four flake-tools were also in the assemblage, including two scrapers, a borer, and a possible burin. Raw material, flake size, and morphology suggest that this site represents the African later MSA.

GnJm 23 (Kipsing): Lithic Scatter, Multiple Component (MSA, LSA, IRA, Historic)

This site is located on the eroding slopes of the south bank of the Kipsing River just east of its junction with the Seaku River. The site continues along the bank of the Kipsing River from its junction with the Seaku River at least several hundred meters downstream. The site is also expressed on the opposite bank. The site stretches inland from each bank about 50-75 m. The site occurs on the Kipsing surface.

Artifact density is low but consistent over the entire area. This is the location of contemporary Kipsing town, which consists of about 50 Turkana huts and associated structures (mainly corrals), and six small wooden buildings. The Kipsing Academy and its walled compound are off the site to the south, while Kenya's General Service Unit (GSU) has occupied several brick and mortar buildings on the north side of the Kipsing River that were built and owned by Kenya's Livestock Marketing Division. This site has been the locus of constant activity in historical times, having been an important point of operation for the British.

We partially surface-collected the prehistoric record, including 53 lithic artifacts. Of these, 60.38 percent (n=32) are basalt, 33.96 percent (n=18) are chert, and 5.66 percent (n=3) are obsidian. The average flake size is 35.66 mm long, 33.31 mm wide, and 11.90 mm thick. Of the 53 artifacts, 18.87 percent (n=10) are basalt cores, 3.77 percent (n=2) are chert cores, and 1.89 percent (n=1) are obsidian cores. Basalt cores were flaked in a radial pattern, while the chert cores were bidirectionally flaked, and the obsidian was flaked "randomly." Five core-tools were observed, as well as four flake-tools, including a chert knife, basalt scrapers, and a backed knife. The variety of material types and technologies represented point to the fact that this location has been occupied through time by a variety of cultures. The Kipsing River is incised into bedrock here and has been fixed at this position for some time. Thus it appears that this location at the confluence of the Kipsing, Seaku, and Tol rivers has been an important locus of human activity from at least the Middle Stone Age to present.

GnJm 24: Stone Cairns

Three stone cairns that make up this site are located on a low rise above the south bank of the Kipsing River downstream from its junction with the Seaku. Coincidentally or not, the site "points to" a cone-shaped mountain due north across the river. All three stone cairns are within 6 m of each other. The site occurs on the Kipsing surface. Kipsing town is about 700 m to the southwest (bearing 246°).

GnJm 25: Stone Cairns, Lithic Scatter, Middle Stone Age and Later

This site consists of two stone cairns and associated lithic material, located on the east bank of the Seaku River on the Peleta surface. The nearest site is GnJm 12, located approximately 750 m to the northwest (bearing 333°). The larger of the two cairns is about 4 m in diameter, while the smaller is about 2 m in diameter. They were separated by about 9 m.

Twelve artifacts were collected from the site. Of these, 75.00 percent (n=9) are basalt and 25.00 percent (n=3) are chert. The average flake size is 35.59 mm long, 32.67 mm wide, and 8.27 mm thick. Of the 12 artifacts, 8.33 percent (n=1) are radially flaked basalt cores. Additionally, one core-tool, a chopper, was also in the assemblage. The crude lithic material associated with the cairns is indicative of the MSA, but we generally believe the cairns are much more recent.

GnJm 26: Stone Cairn

This site is located next to a large outcrop of bedrock along the east bank of the Seaku River on the Peleta surface. The site is a single stone cairn, roughly rectangular, measuring 2 m by 1.5 m. The nearest archaeological site is GnJm 14, which is about 225 m to the southwest (bearing 226°).

SITES ENCOUNTERED DURING THE 1999 SURVEY

GnJm 27: Ceramic Scatter, Iron Age

This site is located on the gentle slope on the east side of Kakwa Lelash. It consists of a single dense concentration of plain ceramic sherds (n > 100), with some other sherds scattered within 20 m. One small piece of copper was also noted here. It is approximately 500 m west of the Tol River. The nearest archaeological site is GnJm 28, located about 245 m to the west (bearing 262°). The surface sediments here are complex. Colluvium from Kakwa Lelash interfingers with Tol sediments.

GnJm 28: Ceramic Scatter, Iron Age

This site is located on the gentle slope on the east side of Kakwa Lelash. This site consists of just a few plain ceramic sherds and one small fragment of ostrich eggshell. It is approximately 160 m east of Kakwa Lelash. The nearest archaeological site is GnJm 29, located about 210 m to the west (bearing 258°). The surface sediments here are complex. Colluvium from Kakwa Lelash interfingers with Tol sediments.

GnJm 29: Multiple Component Prehistoric Use Site (MSA, LSA, IRA)

At the foot of Kakwa Lelash is a group of large boulders. The angle of some of these boulders creates an overhang that was a natural windbreak at various times in prehistory. Archaeological material was observed beneath the overhang and scattered nearby. Artifacts observed on the surface include stone artifacts of obsidian and chert, as well as basalt flakes. This site is located right at the base of the inselberg, and directly beneath Kakwa Lelash Rockshelter (GnJm 2) located up on the inselberg. There is at least 30 cm of deposition beneath the overhang. This may be an important site given its proximity to the rockshelter habitation site. The position of GnJm 29 makes it a logical point of egress en route to Kakwa Lelash Rockshelter. It is the last level ground before a 100 m vertical ascent to the rockshelter. The surrounding large granite boulders provide a natural shelter as well as a measure of privacy and security. The nearest archaeological site is GnJm 28, located approximately 210 m to the east (bearing 77°). Most of the deposition here is recent colluvium (Tol) overlying the Peleta unit.

While large basalt debitage suggests an MSA presence, there is abundant evidence of later prehistoric use. The small overhang contains several panels of contemporary artwork in a reddish pigment media apparently contributed by Samburu residents of the area. Walking northward from the site, one finds intermittent microflakes. I would hypothesize that this site has seen continuous use over a long period of time. It is a logical *ad hoc* workplace for occupants of Kakwa Lelash Rockshelter, and might even be considered a locus of that site.

Because of the proximity of this site to GnJm 2 and because it might potentially be excavated in the future as part of the larger pattern of research associated with that site, only 12 fragmentary lithic specimens were collected from GnJm 29. However, the variety of material types is reflected in this small sample. Of the artifacts collected, 42.67 percent (n=5) are chert, 42.67 percent are quartz, and 16.67 percent (n=2) are obsidian. Basalt artifacts were observed but not collected. No tools or cores were observed. A few plain ceramic sherds were also noted beneath the overhang.

GnJm 30: Ceramic and Lithic Scatter, Later Stone Age

This site is a localized (25m^2) scatter of plain ceramics and obsidian flakes located near the southeastern foot of Kakwa Lelash on the Peleta surface. The nearest archaeological site is GnJm 32, approximately 65 m to the south (bearing 190°).

GnJm 31: Lithic Scatter, Middle Stone Age

This site is a very large (about 4000m^2) lithic scatter about 150 m south of Kakwa Lelash on the Peleta surface. Material appears to have been eroded and redeposited in its current location along the shallow runoff gullies leading away from the Kakwa Lelash inselberg. However, there was no site upslope from which these materials could have been derived. The slope here is about 3 percent. This is an extremely dense site covering a large area, suggesting that it is deflated but perhaps not displaced much. GnJm 32 is the nearest archaeological site and is located 65 m to the north (bearing 10°).

Eighty-five lithic artifacts were collected from the site. Of these, 91.76 percent (n=78) are basalt, 4.71 percent (n=4) are chert, 1.18 percent (n=1) are obsidian, and 2.35 percent (n=2) are quartz. The average flake size is 42.96 mm long, 41.58 mm wide, and 12.98 mm thick. Of the 85 artifacts, 7.06 percent (n=6) are radially flaked basalt cores. Additionally, two core-tools exhibit some battering. Two flake-tools were also in the assemblage, a denticulated basalt knife and a borer. Additionally, one hammerstone was recovered. A single ceramic sherd was found in the vicinity, but is not conclusively associated with the site. This assemblage (not including the sherd) appears to be MSA.

GnJm 32: Lithic Scatter, Middle and Later Stone Age

This site is located on the badly eroding Peleta surface between the foot of Kakwa Lelash and the Tol River. This site is about 50 m south of Kakwa Lelash. Material is scattered around the erosional ditches created by large-scale landscape change. The nearest archaeological site is GnJm 31, located approximately 65 m to the south (bearing 190°).

Fifty-four artifacts were collected from the site. Of these, 9.26 percent (n=5) are basalt, 50.00 percent (n=27) are chert, 35.19 percent (n=19) are obsidian, and 5.56 percent (n=3) are quartz. The average flake size is 24.82 mm long, 21.92 mm wide, and 6.37 mm thick. Of the 54 artifacts, 3.70 percent (n=2) are unidirectionally flaked basalt cores, one exhibiting some battering (classified as a chopper). Also, one pressure-flaked chert microlithic flake-tool rounds out the assemblage. Material types, as well as flake and tool attributes indicate that this site is LSA.

GnJm 33: Lithic Scatter

Two basalt artifacts were collected from the site located in the bottom of a sandy gully leading to the Tol River – undoubtedly redeposited. There is not enough material here to reliably estimate an age. The nearest other site is GnJm 0/10, approximately 190 m to the northeast (bearing 28°).

GnJm 34: Lithic Scatter, Middle Stone Age

This site is located on a high terrace above the Kipsing River, about 210 m south of its junction with the Seaku. This site is on the Kipsing surface. The nearest archaeological site is GnJm 22, located approximately 375 m to the southwest (bearing 240°).

Twenty-two artifacts were collected from the site. All of these are basalt. There were no complete flakes, but almost half the assemblage was represented by cores or tools. Of the 22 artifacts, 22.73 percent (n=5) are basalt cores. Two additional cores show usewear (9.09 percent). Also, 9.09 percent (n=2) are flake tools (end-scrapers). These tools and cores appear to be MSA.

GnJm 35: Ceramic Scatter, Iron Age

These ceramics were located near the west bank of the Tol River, south of Kakwa Lelash. They were found on the Peleta surface and are currently subjected to extreme sheet erosion. No lithic material was found. The Tol River is approximately 280 m to the southeast. The nearest archaeological site is GnJm 39, approximately 225 m to the south (bearing 186°).

GnJm 36: Lithic Scatter, Middle Stone Age

The majority of lithics from this site appear to be MSA. They were recovered on the eroding edge of the first terrace on the east side of the Tol River. They are associated with channel gravels beneath the terrace, indicating that either the site represents a deflationary lag on top of the gravels, or that the site was transported here along with the gravels. As the artifacts are not particularly water-worn and there is little sorting by size, it appears that lag deposition is more likely. This is an example of the gravelly calcic horizon of the Peleta soil. It was eroded to this more resistant layer and the MSA artifacts were left upon its surface as a lag deposit. Subsequently, the site was buried by Shordika alluvium. The site is now being re-exposed by erosion. The Tol River is about 50 m to the west. The nearest other archaeological site is GnJm 42, approximately 125 m to the southwest (bearing 210°).

Seventy-three artifacts were collected from the site. Of these, 83.56 percent (n=61) are basalt, 5.48 percent (n=4) are chert, 2.74 percent (n=2) are obsidian, and 8.22 percent (n=6) are quartz. The average flake size is 32.09 mm long, 36.07 mm wide, and 10.84 mm thick. Of the 73 artifacts, 1.37 percent (n=1) are radially flaked basalt cores. An additional core (classified as a chopper) exhibits some use.

GnJm 37: Lithic Scatter, Middle Stone Age

Bedrock at this site is very shallow and outcrops to the surface where this flake scatter was found. The bedrock has acted as a sediment trap, protecting a portion of the Peleta alluvium from erosion. Lenkiteng is less than 100 m to the east. The Lenkiteng site, GnJm 38) is 145 m to the south (bearing 174°).

Twenty-six basalt artifacts were collected from the site. The average flake size is 40.41 mm long, 45.35 mm wide, and 11.94 mm thick. Of the 26 artifacts, 19.23 percent (n=5) are basalt cores (four radially flaked, one randomly flaked). The material types and flake attributes indicate that this site is MSA.

GnJm 38 (Lenkiteng Site): Buried Lithic Scatter, Later Stone Age

This site is buried in the cut bank of a small gullied-tributary of the Tol River that the local people call *Lenkiteng* (Cattle Creek). The bank is about 3 m deep, in a gully that skirts the eastern edge of a large low-relief granite boulder. The nearest other archaeological site is GnJm 37, approximately 145 m to the north (bearing 344°). Examination of the natural cut bank of Lenkiteng shows that the site is buried in a sequence that includes both Tol and Seaku alluvium over bedrock.

The site occurs on top of a gravelly horizon beneath about 1.5 m of sandy loam. The gravel lens is permeated with carbonates. The site is either in situ or deflated because: 1) the sorting is too poor to be fluvial, and 2) there is little rounding if any on these artifacts (some small flakes still have a sharp edge).

During survey, we recovered 34 artifacts from the site. Of these, 29.41 percent (n=10) are basalt, 38.24 percent (n=13) are chert, 11.76 percent (n=4) are obsidian, and 20.59 percent (n=7) are quartz. The average flake size is 35.84 mm long, 30.75 mm wide, and 9.24 mm thick. Even though complete flake size was large, most small flaking debris was nonlocal material, and chert was the dominant raw material type.

Archaeological investigations began at GnJm 38 in August of 2000. A single 2.0 m x 1.0 m *sondage* was excavated at the natural cut bank where the site was known to exist. This single test unit was excavated down to bedrock. Examination of artifacts recovered from this excavation is currently underway. Hundreds of lithic specimens were recovered. There were high frequencies of exotic source materials, such as chert and obsidian, as well as abundant use of local material. Micro-blades and microflakes were very common, but only one core was recovered. Generally, lithic manufacturing technique suggests an LSA site.

Several geologic profiles were also excavated here (profiles 10a and 10b in Appendix B).

GnJm 39: Ceramic Scatter, Iron Age

A few ceramic sherds were discovered on the dissected Peleta surface west of the upper Tol River just south of the boundary between Laikipia and Isiolo Districts. The Tol River is about 210 m to the east. The nearest archaeological site is GnJm 35, approximately 220 m to the north (bearing 6°).

GnJm 40: Lithic Scatter, Middle Stone Age

These MSA flakes were recovered from the channel of an erosional gully that flows into the Lenkiteng stream, indicating that these artifacts are being transported downstream. This site demonstrates that you can indeed find MSA sites in a Holocene *erosional* context. This opens the possibility that they could be anywhere within Holocene soils in a secondary context. The nearest archaeological site is the Lenkiteng Site (GnJm 38), located downstream (north) about 155 m (bearing 14°).

Only two basalt artifacts were collected from the site. One, a large basalt flake with medial/lateral denticulation is distinctive of MSA technology.

GnJm 41: Lithic and Ceramic Scatter, Later Stone Age

This site is located between the Tol River and Lenkiteng, on the Shordika surface west of a small inselberg known as "Bumbu." This inselberg is less than 100 m tall and does not appear on the topographic map. Several sherds and obsidian flakes were found in and around a small erosional gully. One small rim sherd was found with a punctate decoration, generally indicative of the LSA. This site is approximately 410 m southeast of the Tol River. The nearest other archaeological site is GnJm 0/15, approximately 160 m to the southwest (bearing 237°).

GnJm 42: Lithic Scatter, Middle Stone Age

This site is on the east side of the Tol River on the Peleta surface. It is exceptional for the number of smaller MSA flakes present. The position on the landscape is close to the Tol in the highly erosional 3 percent slope (heavily dissected). It looks as if the site is eroding from beneath Shordika sediment (similar to the context of GnJm 36). The Tol River is less than 100 m to the west. The nearest archaeological site is GnJm 36, approximately 125 m to the northeast (bearing 30°). Further investigation may reveal that these sites are related.

Sixty lithic artifacts were collected from the site. Of these, 93.33 percent (n=56) are basalt, and 6.67 percent (n=4) are chert. The average flake size is 36.44 mm long, 40.46 mm wide, and 12.60 mm thick. Of the 60 artifacts, 5.00 percent (n=3) are basalt cores. An additional core shows signs of use (classified as a chopper). Local raw materials are dominant in this assemblage, but flake size is intermediary between the Shurmai MSA and Kakwa Lelash collections, so a late MSA date is appropriate.

GnJm 43: Lithic Scatter, Later Stone Age

This site is located on the Shordika surface at the foot of a small inselberg known as "Bumbu." This inselberg is less than 100 m tall and does not appear on the topographic map. Some of the lithic material appears on the rocky surface of the inselberg and has apparently washed downward. Several boulders stand to the east of Bumbu, allowing a good wind-break. Later Stone Age material was found there as well. Lenkiteng (creek) is approximately 400 m to the east. The nearest archaeological site is GnJm 44, approximately 50 m to the south (Bearing 180°).

Sixty-four lithic artifacts were collected from the site. Of these, 70.31 percent (n=45) are basalt, 3.13 percent (n=2) are chert, 4.69 percent (n=3) are obsidian, and 21.88 percent (n=14) are quartz. The average complete flake size is 29.04 mm long, 39.49 mm wide, and 13.02 mm thick. One obsidian microlithic end-scraper is among the 64 artifacts. Smaller flake size and microlithic industry place this site squarely in the LSA.

GnJm 44: Lithic and Ceramic Scatter, Later Stone Age

This site is a small scatter of obsidian flakes and ceramic sherds directly on the Shordika surface south of a small inselberg known as "Bumbu." This inselberg is less than 100 m tall and does not appear on the topographic map. Lenkiteng (creek) is approximately 400 m to the east. The nearest archaeological site is GnJm 43, approximately 50 m to the north (Bearing 360°).

Eighteen lithic artifacts were collected from the site. Of these, 5.56 percent (n=1) are basalt, 83.33 percent (n=15) are obsidian, and 11.11 percent (n=2) are quartz. There are too few complete flakes to calculate a reliable flake size, but flaking debris is generally indicative of microlithic technology. No tools were present. The presence of ceramics also indicates an LSA date for the site.

GnJm 46: Lithic and Ceramic Scatter, Later Stone Age

This site is located north of Peleta Rock immediately east of the road that runs between Peleta Creek and the Lenkiteng. The site is eroding out of the Shordika surface at the boundary between the modern floodplain and the first terrace. The site is approximately 300 m west of Peleta Creek. The nearest archaeological site is GnJm 51, located approximately 300 m to the east (bearing 90°).

In addition to some scattered plain ceramic sherds, one pressure-flaked obsidian side-scraper was recovered. These artifacts suggest an LSA date for the site.

GnJm 47 (Peleta Rock): Lithic and Ceramic Scatter, Multiple Component (MSA, LSA, Historic)

Peleta Rock is a prominent granitic inselberg near the entrance of the Shurmai Valley, which leads to GnJm 1,

Shurmai Rockshelter. It is several hundred meters long and about 100 meters tall. It is often the locus of local activity as Samburu *moran* occasionally meet there in small numbers. Locals often refer to gunshots heard in the distance as coming from the *moran* at Peleta Rock. The rock itself is large and flat enough on top that we observed someone walking up there on more than one occasion. It is a convenient geographic marker and is often referred to by people travelling through the area. The British were known to have erected a radio antenna here and to have excavated a water pond on the south side of the rock (see GnJm 48). One informant said that the British camped up on Peleta Rock to avoid elephants down below. However, there are no visible signs of a British camp up on the rock. There is preserved sediment on a rocky ledge on the western side that is wind sheltered and could contain cultural material of historic and prehistoric significance.

The survey was principally of the surface around Peleta, and abundant lithic material was discovered on the eastern side. It appears that the Peleta surface underlies Shordika sediments here, but sheet erosion has reveal portions of the Peleta unit. There may be some intermixing of the two units here. Peleta Creek is about 200 meters east of Peleta Rock. The nearest other archaeological site is GnJm 48, approximately 180 m southwest of where these artifacts were discovered (bearing 234°).

One hundred and eleven lithic artifacts were collected from the east side of the rock. Of these, 41.44 percent (n=46) are basalt, 25.23 percent (n=28) are chert, 29.73 percent (n=33) are obsidian, and 3.60 percent (n=4) are quartz. The average complete flake size is 32.30 mm long, 25.52 mm wide, and 6.67 mm thick. Surprisingly few tools were seen in this collection. The only flake tool is an obsidian flake with use wear along the proximal end. Two basalt core-tools showed some usewear as well. A few ceramics were found on the eastern edge of the site (about 60 m from Peleta Rock). The site clearly has an LSA component, but probably many of the larger basalt flakes and core-tools are MSA material. Peleta Rock has undoubtedly been visited from the later Pleistocene to present.

GnJm 48: Lithic Scatter, Later Stone Age and Historic

This site is on the northern edge of Peleta Rock. It is distinguished from GnJm 47 only because of its tight boundaries. There is a small natural protected area created by the coming together of Peleta Rock and several large boulders. One informant suggested that this area was excavated by the British to form a collection pond for water. Obsidian artifacts were observed on the surface within the windbreak as well as on the surface away from the area, so it is unclear how much may have been disturbed by recent reworking. Probing detected that there are about 70 cm of deposits here. Peleta Creek is about 200 meters east of Peleta Rock. The nearest other archaeological site is GnJm 47, approximately 180 m northeast of where these artifacts were discovered (bearing 54°). Four obsidian microlithic flakes were collected from the site.

GnJm 49: Lithic and Ceramic Scatter, Later Stone Age

This is a small lithic and ceramic scatter on the Shordika surface. A single obsidian tool and a small collection of plain ceramic sherds indicate that this site is LSA. Lenkiteng is about 80 m to the west. The nearest archaeological site is GnJm 56, located approximately 290 m to the southeast (bearing 148°).

GnJm 50: Lithic Scatter (Middle Stone Age)

This site is located on an eroding side-slope of exposed gravels of the Peleta unit underlying the Shordika surface as it approaches the Tol River. This is the same geomorphic context as GnJm 36. The Tol River is less than 100 m to the west. The nearest archaeological site is GnJm 0/15, located about 350 m to the east (bearing 72°).

Ten lithic artifacts were collected from the site. Of these, 80.00 percent (n=8) are basalt and 10.00 percent (n=1) are chert. There are too few artifacts for these statistics to be meaningful. Its position on the landscape, however, suggests that it is an MSA site.

GnJm 51: Lithic Scatter, Middle Stone Age (Late)

This lithic scatter is on the east bank of Peleta Creek on the Peleta surface. The nearest archaeological site is GnJm 52, located approximately 170 m to the southeast (Bearing 139°). Lithics at this site were entirely of locally available raw material.

Thirty-eight basalt artifacts were collected from the site. The average complete flake size is 34.72 mm long, 33.11 mm wide, and 9.96 mm thick. Two cores were among the assemblage. Flake size is somewhat smaller than the MSA assemblage from Shurmai Rockshelter, suggesting that this site is probably late MSA.

GnJm 52: Ceramic Scatter, Iron Age

Numerous decorated (slipped) pottery and rim sherds were collected from this site. These slippedwares indicate a post-lithic date for the site. The site is situated on the Peleta surface, about 25 m east of Peleta Creek. The nearest other archaeological site is GnJm 0/17, located approximately 85 m to the west (bearing 252°).

GnJm 53 (Mau-Mau Camp): Historic Site

According to a local Mukogodo informant, during the Mau-Mau rebellion, the Mukogodo had a large *manyatta* here in the foothills of Shordika. The Mau-Mau rebellion was a guerilla-style uprising against the British colonial government, principally by Kikuyu, that began in earnest in 1953. Though it was ostensibly unsuccessful, many believe that it contributed to the social movement that led to Kenyan independence in 1962.

According to our informant, some Kikuyu rebels came here and made an agreement with the Mukogodo to fight on their side. Our informant was a *moran* during that time and remembers taking the Mau-Mau oath. It should be noted that our informant was speaking to us in Swahili (his second or third language) through a Kikuyu interpreter.

According to this informant, the Mukogodo had been employed by the British to keep the forest Mau-Mau free, and were given a bolt-action Ensfield rifle equipped with bayonet to do his job. However, when the Mau-Mau came he was either forced or coerced to take the oath. The British did indeed hire the Mukogodo and their neighbors to search for "terrorists" in the 1950s (Cronk 1989a:115). One local man even caught a Mau-Mau rebel single handedly. The British also built a radio station near Peleta Rock (see GnJm 47) to monitor Mau-Mau activity in this area. If this story is correct, the Mukogodo were pawns in this political game.

Aside from a few pieces of tin here and there and some stones marking a cooking hearth, very little material was observed at the site. Some pioneer vegetation marks the site. The site is on the lower slopes of Shordika Hill where Shurmai Rockshelter (GnJm 1) is located. It is approximately 400 m east of Peleta Creek on the sloping Peleta surface. The nearest archaeological site is GnJm 0/19, located approximately 430 m to the southwest (bearing 247°).

GnJm 54: Lithic Scatter, Middle Stone Age

This site is a very broad lithic scatter on the crest of the Shurmai Valley. It is southwest of and beneath Shurmai Rockshelter (GnJm 1). Only 400 m away from Shurmai Rockshelter, it is the nearest Paleolithic find to that site. However, the closest archaeological site to GnJm 54 is GnJm 55, located approximately 150 m to the southeast (bearing 126°). Sediments here are Peleta colluvium.

Fifteen lithic artifacts were collected from the site. The average complete flake size is 46.17 mm long, 41.78 mm wide, and 12.29 mm thick. Three basalt cores are in the assemblage (20.00 percent), as well as one long core-tool (128.07 mm) showing end battering from use. These implements are large and are as old, or older, than the MSA in Shurmai.

GnJm 55: Iron Smelting Site, Iron Age

This site consists of a broad scatter of corroding iron slag, some ceramic sherds, and several iron-smelting kilns. Each hearth has eroded down to the ground and not much is left of them. Each consists of about a 0.5 meter diameter outer kiln wall, with slag and burned earth lying on the inner surface. Remains of a ceramic vent pipe lead away from one kiln. There is no water onsite, but natural cisterns are found on Shordika Hill in Shurmai Rockshelter (GnJm 1). The stratigraphic position of the site is on the Peleta surface. The nearest archaeological site is GnJm 54, located approximately 150 m to the northwest (bearing 306°).

GnJm 56 (Police Post): Historic Site

This site dates to around the time of Kenya's independence. According to a local informant, a small police post was built here to protect the locals from Somalian banditry. It was surrounded by a ditch and low earthen wall about 100 m in diameter. The ditch is still visible, as well as a deep overgrown trash pit. The location of the former building is marked by relict wooden posts protruding from the ground. The remnants of a cement platform at the site was said to have been built for the radio. Our informant notes that locals were employed in the construction of the facility but were paid only with food as wages.

The stratigraphic position of the site is on the Shordika surface. The nearest other archaeological site is GnJm 49, located approximately 290 m to the northwest (bearing 329°).

GnJm 57: Lithic Scatter, Middle Stone Age

This small lithic scatter is located in the Shurmai Valley east of Peleta Rock. The nearest archaeological site is GnJm 48 (at Peleta Rock), located across Peleta Creek approximately 455 m to the northwest (bearing 344°). The lithic material was eroding out of the Peleta surface.

Ten lithic artifacts were collected from the site. Of these, 90.00 percent (n=9) are basalt, and 10.00 percent (n=1) are obsidian. The two complete flakes are large (over 50 mm long), and a large core tool was also recovered. These large artifacts are characteristic of an MSA assemblage.

GnJm 58: Buried Lithic Scatter, Middle Stone Age

This site is located in a deep erosional ditch in the flat, eroded Peleta surface, approximately 1.25 km west of the Tol River. Lithic material appears at a depth of 1 m or more atop the thick calcic horizon in the Peleta soil. Erosion has exposed the calcic carbonate layer and this site. Granitic bedrock is very close to the surface and is decomposing in situ to grüs (decomosing granite) about 1.0 to 1.5 m deep. Red sediment overburden is coming from upstream. The geomorphic context of the site is significant because it is the only known buried MSA site. It overlies the soil carbonates. The nearest other archaeological site is GnJm 60, located approximately 450 m to the northeast (bearing 55°).

Eighteen basalt artifacts were collected from the site. The average complete flake size is 58.09 mm long, 54.20 mm wide, and 15.05 mm thick. Two cores were recovered, and two core-tools which showed some signs of use as a tool. Raw material use, flake size, and tool type are all indicative of the MSA.

GnJm 59: Lithic Scatter, Later Stone Age

This site is located on an eroding slope near a shallow gully-tributary of the Tol River. The site is located on the

Kipsing surface where it overlies the Peleta unit. The Tol River is about 1.25 km to the southeast. The nearest archaeological site is GnJm 60, located approximately 265 m to the northwest (bearing 330°).

Fifteen lithic artifacts were collected from the site. Of these, 33.33 percent (n=5) are basalt, and 66.66 percent (n=10) are chert. The average complete flake size is 23.00 mm long, 20.48 mm wide, and 6.44 mm thick. There were no tools or cores in the assemblage, but the material use and flaking type strongly suggests LSA.

GnJm 60: Lithic Scatter

This site is located on the Kipsing surface, approximately 1.2 km northwest of the Tol River. Two basalt artifacts were collected from the site. They are similar in size and morphology to other MSA artifacts, but the sample is too small to say with certainty. The nearest other archaeological site is GnJm 59, located approximately 265 m to the southeast (bearing 150°).

GnJm 61: Lithic Scatter

This site is located on the Kipsing surface, approximately 2.9 km west of the junction of the Tol and Seaku Rivers. Three lithic artifacts were collected from the site. Two are basalt and one is obsidian. There are no tools or cores, and the sample here is too small to venture an age estimate. The nearest other archaeological site is GnJm 60, located approximately 1.22 km to the south (bearing 190°).

GnJm 62: Lithic Scatter, Middle Stone Age

This site is located on the Kipsing surface less than 25 m from the west bank of the Tol River. Eighteen basalt artifacts were collected from the site. The nearest other archaeological site is GnJm 63, located about 130 m to the northeast (bearing 11°).

The average complete flake size is 36.06 mm long, 35.04 mm wide, and 89.83 mm thick. These characteristics of flake size and material use are nearly identical to the MSA component at Shurmai rockshelter.

GnJm 63: Ceramic Scatter, Iron Age

This site is located on the Kipsing surface, less than 25 m from the west bank of the Tol River. Several rim sherds with incised decorations were collected. This type of sherd and the absence of lithics suggest an Iron Age date. The nearest other archaeological site is GnJm 62, located about 130 m to the southwest (bearing 191°).

GnJm 64: Lithic Scatter, Middle Stone Age

This site is located on the Kipsing surface, about 50 m from the west bank of the Tol River. Thirty-eight lithic artifacts were collected from the site. Of these, 97.37 percent (n=37) are basalt and 2.63 percent (n=1) are quartz. The average complete flake size is 58.40 mm long, 57.45 mm wide, and 19.82 mm thick. There were no tools or cores in the assemblage, but material use and large flake size strongly suggest MSA. The nearest archaeological site is GnJm 65, located approximately 470 m to the northeast (bearing 24°).

GnJm 65: Lithic Scatter, Middle Stone Age

This site is located on the Kipsing surface, about 500 m west of the Tol River. Eight lithic artifacts were collected from the site. Of these, 87.50 percent (n=7) are basalt and 12.50 percent (n=1) are chert. The average complete flake size is 39.29 mm long, 34.45 mm wide, and 11.99 mm thick. There is one large (62.55 mm long, 66.15 mm wide, and 25.47 mm thick) flake tool in the assemblage with edge modification. This tool type, and the overall size of the assemblage, strongly suggests an MSA date. The nearest other archaeological site is GnJm 64, approximately 470 m to the southwest (bearing 204°).

GnJm 0/10: Isolated Lithic Flake

A single basalt flake was collected from this site located on the Peleta surface. There is not enough material here to reliably estimate an age. The nearest other site is GnJm 33, approximately 190 m to the southwest (bearing 208°).

GnJm 0/11: Isolated Lithic Flake

A single basalt flake was located in Tol age sediments in the Lenkiteng channel. There is no attributable age. The nearest archaeological site is GnJm 0/12, located approximately 265 m to the southwest (bearing 205°).

GnJm 0/12: Isolated Lithic Scatter

Six basalt artifacts and a single quartz core were collected from Tol age sediments of the Lenkiteng channel bottom. It is difficult to interpret these finds because they are incorporated into the channel gravels, having been well sorted and transported from elsewhere. They have no assignable age. The nearest archaeological site is GnJm 0/11, located approximately 265 m to the northeast (bearing 24°).

GnJm 0/13: Ceramic Scatter, Iron Age

Several plain IRA ceramic sherds with no lithic material were collected from this site on the Peleta surface. The Tol River is approximately 430 m to the east. The nearest archaeological site is GnJm 35, approximately 300 m to the southwest (bearing 223°).

GnJm 0/14: Ceramic Scatter, Iron Age

Several plain IRA ceramic sherds with no lithic material were collected from this site on the Shordika surface. The Tol River is approximately 320 m to the west. The nearest archaeological site is GnJm 42, approximately 210 m to the northeast (bearing 16°).

GnJm 0/15: Isolated Lithic

A single basalt fragment was discovered at this site on the Shordika surface. The Tol River is approximately 470 m to the west. The nearest archaeological site is GnJm 41, approximately 160 m to the northeast (bearing 58°)

GnJm 0/16: Isolated Lithics

One core and one core tool were collected from this site on the lower slopes of Shordika Hill (where Shurmai Rockshelter [GnJm 1] is located). It is approximately 580 m east of Peleta Creek on the sloping Peleta surface. The nearest archaeological site is GnJm 53, located approximately 930 m to the southwest (bearing 220°). These artifacts have no attributable age.

GnJm 0/17: Isolated Lithic Scatter

Six basalt artifacts and a single quartz core were collected from a gully adjacent to Peleta Creek. It is difficult to interpret these finds because they are incorporated into the channel gravels, having been well sorted and transported from upslope on Shordika Hill. They have no attributable age. The nearest archaeological site is GnJm 52, approximately 75 m to the southwest (bearing 248°).

GnJm 0/18: Lithic and Ceramic Scatter, Later Stone Age

This site consists of a single piece of obsidian degitage and a small collection of plain ceramic sherds, indicating that this site is attributable to the LSA. The site is situated on the Shordika surface, approximately 380 m west of Lenkiteng. The nearest archaeological site is GnJm 49, located about 460 m to the east across Lenkiteng (bearing 103°).

GnJm 0/19: Ceramic Scatter, Iron Age

This site is a small collection of undecorated ceramic sherds attributable to the Iron Age. The site is located on the Shordika surface just above the west bank of Peleta Creek. The nearest archaeological site is GnJm 47, approximately 260 m to the southwest (bearing 244°).

GnJm 0/20: Ceramic Scatter, Iron Age

This site is a small collection of undecorated ceramic sherds attributable to the Iron Age. The site is located on the Kipsing surface approximately 525 m west of the Tol River. The nearest archaeological site is GnJm 63, approximately 475 m to the southeast (bearing 108°).

GnJm 0/21: Ceramic Scatter, Iron Age

This site is a small collection of undecorated ceramic sherds attributable to the Iron Age. It is located on the Shordika surface about 100 m west of the Tol River. The nearest archaeological site is GnJm 47, approximately 350 m to the northeast (bearing 64°).

GnJm 0/22: Isolated Lithic Flake

This find consists of a single basalt flake embedded in Tol sediments in a cutbank of the Tol River. It is noteworthy because it demonstrates the possibility of buried sites beneath the Tol surface, but it has no attributable age. The nearest archaeological site is GnJm 62, located approximately 250 m due north (bearing 360°).

CAMBRIDGE MONOGRAPHS IN AFRICAN ARCHAEOLOGY

No 1   BAR S75, 1980 **The Niger Delta** *Aspects of its Prehistoric Economy and Culture* by Nwanna Nzewunwa. ISBN 0 86054 083 9

No 2   BAR S89, 1980 **Prehistoric Investigations in the Region of Jenne, Mali** *A Study in the Development of Urbanism in the Sahel* by Susan Keech McIntosh and Roderick J. McIntosh ISBN 0 86054 103 7

No 3   BAR S97, 1981 **Off-Site Archaeology and Human Adaptation in Eastern Africa** *An Analysis of Regional Artefact Density in the Amboseli, Southern Kenya* by Robert Foley. ISBN 0 86054 114 2

No 4   BAR S114, 1981 **Later Pleistocene Cultural Adaptations in Sudanese Nubia** by Yousif Mukhtar el Amin. ISBN 0 86054 134 7

No 5   BAR S119, 1981 **Settlement Patterns in the Iron Age of Zululand** *An Ecological Interpretation* by Martin Hall. ISBN 0 86054 143 6

No 6   BAR S139, 1982 **The Neolithic Period in the Sudan, c. 6000-2500 B.C.** by Abbas S. Mohammed-Ali. ISBN 0 86054 170 3

No 7   BAR S195, 1984 **History and Ethnoarchaeology in Eastern Nigeria** *A Study of Igbo-Igala relations with special reference to the Anambra Valley* by Philip Adigwe Oguagha and Alex Ikechukwu Okpoko. ISBN 0 86054 249 1

No 8   BAR S197, 1984 **Meroitic Settlement in the Central Sudan** *An Analysis of Sites in the Nile Valley and the Western Butana* by Khidir Abdelkarim Ahmed. ISBN 0 86054 252 1

No 9   BAR S201, 1984 **Economy and Technology in the Late Stone Age of Southern Natal** by Charles Cable. ISBN 0 86054 258 0

No 10   BAR S207, 1984 **Frontiers** *Southern African Archaeology Today* edited by M. Hall, G. Avery, D.M. Avery, M.L. Wilson and A.J.B. Humphreys. ISBN 0 86054 268 8. £23.00.

No 11   BAR S215, 1984 **Archaeology and History in Southern Nigeria** *The ancient linear earthworks of Benin and Ishan* by P.J. Darling. ISBN 0 86054 275 0

No 12   BAR S213, 1984 **The Later Stone Age of Southernmost Africa** by Janette Deacon. ISBN 0 86054 276 9

No 13   BAR S254, 1985 **Fisher-Hunters and Neolithic Pastoralists in East Turkana, Kenya** by John Webster Barthelme. ISBN 0 86054 325 0

No 14   BAR S285, 1986 **The Archaeology of Central Darfur (Sudan) in the 1st Millennium A.D.** by Ibrahim Musa Mohammed. ISBN 0 86054 367 6.

No 15   BAR S293, 1986 **Stable Carbon Isotopes and Prehistoric Diets in the South-Western Cape Province, South Africa** by Judith Sealy. ISBN 0 86054 376 5.

No 16   BAR S318, 1986 **L'art rupestre préhistorique des massifs centraux sahariens** by Alfred Muzzolini.. ISBN 0 86054 406 0

No 17   BAR S321, 1987 **Spheriods and Battered Stones in the African Early and Middle Stone Age** by Pamela R. Willoughby. ISBN 0 86054 410 9

No 18   BAR S338, 1987 **The Royal Crowns of Kush** *A study in Middle Nile Valley regalia and iconography in the 1st millennia B.C. and A.D.* by Lázló Török.. ISBN 0 86054 432 X

No 19   BAR S339, 1987 **The Later Stone Age of the Drakensberg Range and its Foothills** by H. Opperman. ISBN 0 86054 437 0

No 20   BAR S350, 1987 **Socio-Economic Differentiation in the Neolithic Sudan** by Randi Haaland. ISBN 0 86054 453 2

No 21   BAR S351, 1987 **Later Stone Age Settlement Patterns in the Sandveld of the South-Western Cape Province, South Africa** by Anthony Manhire. ISBN 0 86054 454 0

No 22   BAR S365, 1987 **L'art rupestre du Fezzan septentrional (Libye) Widyan Zreda et Tarut (Wadi esh-Shati)** by Jean-Loïc Le Quellec. ISBN 0 86054 473 7

No 23   BAR S368, 1987 **Archaeology and Environment in the Libyan Sahara** *The excavations in the Tadrart Acacus*, 1978-1983 edited by Barbara E. Barich. ISBN 0 86054 474 5

No 24   BAR S378, 1987 **The Early Farmers of Transkei, Southern Africa Before A.D. 1870** by J.M. Feely. ISBN 0 86054 486 9

No 25   BAR S380, 1987 **Later Stone Age Hunters and Gatherers of the Southern Transvaal** *Social and ecological interpretation* by Lyn Wadley. ISBN 0 86054 492 3

No 26   BAR S405, 1988 **Prehistoric Cultures and Environments in the Late Quaternary of Africa** edited by John Bower and David Lubell. ISBN 0 86054 520 2

No 27   BAR S418, 1988 **Zooarchaeology in the Middle Nile Valley** *A Study of four Neolithic Sites near Khartoum* by Ali Tigani El Mahi. ISBN 0 86054 539 3

No 28   BAR S422, 1988 **L'Ancienne Métallurgie du Fer à Madagascar** by Chantal Radimilahy. ISBN 0 86054 544 X

No 29   BAR S424, 1988 **El Geili The History of a Middle Nile Environment, 7000 B.C.-A.D. 1500** edited by I. Caneva. ISBN 0 86054 548 2

No 30   BAR S445, 1988 **The Ethnoarchaeology of the Zaghawa of Darfur (Sudan) Settlement and Transcience** by Natalie Tobert. ISBN 0 86054 574 1

No 31   BAR S455, 1988 **Shellfish in Prehistoric Diet Elands Bay, S.W. Cape Coast, South Africa** by W.F. Buchanan. ISBN 0 86054 584 9

No 32   BAR S456, 1988 **Houlouf I** *Archéologie des sociétés protohistoriques du Nord-Cameroun* by Augustin Holl. ISBN 0 86054 586 5

No 33   BAR S469, 1989 **The Predynastic Lithic Industries of Upper Egypt** by Liane L. Holmes. ISBN 0 86054 601 2 (two volumes)

No 34   BAR S521, 1989 **Fishing Sites of North and East Africa in the Late Pleistocene and Holocene** *Environmental Change and Human Adaptation* by Kathlyn Moore Stewart. ISBN 0 86054 662 4

No 35   BAR S523, 1989 **Plant Domestication in the Middle Nile Basin** *An Archaeoethnobotanical Case Study* by Anwar Abdel-Magid. ISBN 0 86054 664 0

No 36   BAR S537, 1989 **Archaeology and Settlement in Upper Nubia in the 1st Millennium A.D.** by David N. Edwards. ISBN 0 86054 682 9

No 37   BAR S541, 1989 **Prehistoric Settlement and Subsistence in the Kaduna Valley, Nigeria** by Kolawole David Aiyedun and Thurstan Shaw. ISBN 0 86054 684 5

No 38   BAR S640, 1996 **The Archaeology of the Meroitic State** *New perspectives on its social and political organisation* by David N. Edwards. ISBN 0 86054 825 2

No 39   BAR S647, 1996 **Islam, Archaeology and History** *Gao Region (Mali) ca. AD 900 - 1250* by Timothy Insoll. ISBN 0 86054 832 5

No 40   BAR S651, 1996 **State Formation in Egypt**: *Chronology and society* by Toby A.H. Wilkinson. ISBN 0 86054 838 4

No 41   BAR S680, 1997 **Recherches archéologiques sur la capitale de l'empire de Ghana** *Etude d'un secteur d'habitat à Koumbi Saleh, Mauritanie. Campagnes II-III-IV-V (1975-1976)-(1980-1981)* by S. Berthier. ISBN 0 86054 868 6

No 42   BAR S689, 1998 **The Lower Palaeolithic of the Maghreb** *Excavations and analyses at Ain Hanech, Algeria* by Mohamed Sahnouni. ISBN0 86954 875 9

No 43   BAR S715, 1998 **The Waterberg Plateau in the Northern Province, Republic of South Africa, in the Later Stone Age** by Maria M. Van der Ryst. ISBN 0 86054 893 7

No 44   BAR S734, 1998 **Cultural Succession and Continuity in S.E. Nigeria** *Excavations in Afikpo* by V. Emenike Chikwendu. ISBN 0 86054 921 6

No 45   BAR S763, 1999 **The Emergence of Food Production in Ethiopia** by Tertia Barnett. ISBN 0 86054 971 2

No 46   BAR S768, 1999 **Sociétés préhistoriques et Mégalithes dans le Nord-Ouest de la République Centrafricaine** by Étienne Zangato. ISBN 0 86054 980 1

No 47   BAR S775, 1999 **Ethnohistoric Archaeology of the Mukogodo in North-Central Kenya** *Hunter-gatherer subsistence and the transition to pastoralism in secondary settings* by Kennedy K. Mutundu. ISBN 0 86054 990 9

No 48   BAR S782, 1999 **Échanges et contacts le long du Nil et de la Mer Rouge dans l'époque protohistorique (IIIe et IIe millénaires avant J.-C.)** *Une synthèse préliminaire* by Andrea Manzo. ISBN 1 84171 002 4

No 49   BAR S838, 2000 **Ethno-Archaeology in Jenné, Mali** *Craft and status among smiths, potters and masons* by Adria LaViolette. ISBN 1 84171 043 1

No 50   BAR S860, 2000 **Hunter-Gatherers and Farmers** *An enduring Frontier in the Caledon Valley, South Africa* by Carolyn R. Thorp. ISBN 1 84171 061 X

No 51   BAR S906, 2000 **The Kintampo Complex** *The Late Holocene on the Gambaga Escarpment, Northern Ghana* by Joanna Casey. ISBN 1 84171 202 7

No 52   BAR S964, 2000 **The Middle and Later Stone Ages in the Mukogodo Hills of Central Kenya** *A Comparative Analysis of Lithic Artefacts from Shurmai (GnJm1) and Kakwa Lelash (GnJm2) Rockshelters* by G-Young Gang. ISBN 1 84171 251 5

No 53   BAR S1006, 2001 **Darfur (Sudan) In the Age of Stone Architecture c. 1000 - 1750 AD** *Problems in historical reconstruction* by Andrew James McGregor. ISBN 1 84171 285 X

No 54   BAR S1037, 2002 **Holocene Foragers, Fishers and Herders of Western Kenya** by Karega-Mũnene. ISBN 1 84171 1037

No 55   BAR S1090, 2002 **Archaeology and History in Ìlàrè District (Central Yorubaland, Nigeria) 1200-1900 A.D.** by Akinwumi O. Ogundiran. ISBN 1 84171 468 2

No 56   BAR S1133, 2003 **Ethnoarchaeology in the Zinder Region, Republic of Niger: the site of Kufan Kanawa** by Anne Haour. ISBN 1 84171 506 9

No 57   BAR S1187, 2003 **Le Capsien typique et le Capsien supérieur** *Évolution ou contemporanéité. Les données technologiques* by Noura Rahmani. ISBN 1 84171 553 0

No 58   BAR S1216, 2004 **Fortifications et urbanisation en Afrique orientale** by Stéphane Pradines. ISBN 1 84171 576 X

www.ingramcontent.com/pod-product-compliance
Ingram Content Group UK Ltd.
Pitfield, Milton Keynes, MK11 3LW, UK
UKHW061213180426
11947UKWH00029B/2017